Supplementary Material
and
Solutions Manual
for
Mathematical Modeling in the Environment

ISBN 0-88385-713-8

Printed in the United States of America

Current Printing (last digit):
10 9 8 7 6 5 4 3 2 1

Supplementary Material
and
Solutions Manual
for
Mathematical Modeling in the Environment

Charles R. Hadlock

Published by

THE MATHEMATICAL ASSOCIATION OF AMERICA

CLASSROOM RESOURCE MATERIALS

Classroom Resource Materials is intended to provide supplementary classroom material for students—laboratory exercises, projects, historical information, textbooks with unusual approaches for presenting mathematical ideas, career information, etc.

101 Careers in Mathematics, edited by Andrew Sterrett
Calculus Mysteries and Thrillers, R. Grant Woods
Combinatorics: A Problem Oriented Approach, Daniel A. Marcus
Elementary Mathematical Models, Dan Kalman
Interdisciplinary Lively Application Projects, edited by Chris Arney
Laboratory Experiences in Group Theory, Ellen Maycock Parker
Learn from the Masters, Frank Swetz, John Fauvel, Otto Bekken, Bengt Johansson, and Victor Katz
Mathematical Modeling in the Environment, Charles R. Hadlock
A Primer of Abstract Mathematics, Robert B. Ash
Proofs Without Words, Roger B. Nelsen
A Radical Approach to Real Analysis, David M. Bressoud
She Does Math!, edited by Marla Parker

MAA Service Center
P. O. Box 91112
Washington, DC 20090-1112
1-800-331-1622   fax: 1-301-206-9789

Prologue

This volume contains solutions to all the exercises in the main text as well as additional comments that would primarily be of interest to instructors using this material in their classes. Each chapter is organized and numbered to correspond to the text, and, in fact, each exercise from the text is restated here so that it can be read in a self-contained fashion. To avoid confusion in making references to figures and tables, those introduced in this volume have been lettered instead of numbered. Thus a reference to Figure 2-19, say, would necessarily be to that figure in Chapter 2 of the main text, whereas a reference to Figure 2-D, say, would indicate that the figure is in Chapter 2 of this volume. When additional background or teaching comments are readily separable from the solution to an exercise, they are shown in a distinctive format. This should make it easier for the reader to browse through this volume and identify particular ideas that he or she may want (or may not want) to try. Aside from the more focused comments and advice found throughout this volume, some common themes and general ideas are summarized in this section. One may also want to review the general comments sections at the ends of the individual chapters in the text, as these often have ideas for activities that may also benefit classes using this material.

Meeting the students—geographic ties

When I meet the class for the first time and am trying to learn their names, I ask each student to tell the group the name of his or her hometown. I always hope I know something about a hot environmental issue in at least some of the towns or states or regions (or foreign countries); but even if I don't, I try to work the geographic theme in further as the course proceeds. For example, I might research some of their hometowns and then be able to work them in during a subsequent class, which does really help to make the real world connection that I think is so important.

Field trips

I would encourage instructors to arrange one or more field trips so that students can experience the environmental issues in their real world setting. Students in my own classes invariably say that these are one of the most valuable aspects of the course. In fact, I generally do four a semester—during class time. This entails my teaching the course on a special schedule outside the college's usual format, namely, one three hour block on one day of each week. A colleague has even been able to organize field trips when the course is taught in our evening division and the trips must be at night. Of course, the intensity of the field program limits the time available for class meetings, and there are obvious trade-offs. At the same time, even one trip off-campus to talk with people actually working on environmental problems can be invaluable. For example, few people realize that their local fire department may be using some of the same computer models that are treated in Chapter 4, but it all becomes extremely real when the firemen start talking about them. The practical scheduling constraint of holding field trips during class time may not apply at some other institutions, and this may expand the opportunities.

Good possibilities for field trips include: contaminated hazardous waste sites, especially "Superfund" sites (ground-water contamination), power plants (air pollution), fire departments, especially hazardous materials units (modeling of hazardous materials emergencies), and industrial plants (a full range of such issues). Government environmental agencies can be a good starting point in identifying such opportunities, and there is usually no shortage of industry and government specialists willing to serve as hosts for such field trips.

Some students have occasionally even asked to bring friends from our school or from other schools on the field trips.

When I first started setting up sites for such field trips, I made a number of preliminary visits to various kinds of facilities. In each case, I invited several faculty members from my own and other departments to come along, and this turned out to be a valuable learning and collegial experience.

Naturally, it's best if you can lead up to a field trip by some discussion in class, and follow it up by some further discussion. For example, if going to a power plant, one might first review the basic flow diagram for such a plant. (One is presented in Section 3-8 of the text.) It is also sometimes possible to construct a homework problem around a field trip, or some short-answer questions for an exam that might encourage the students to review some of the things they encountered on such trips.

Interdisciplinary orientation

Don't feel that you're being fraudulent by getting up in front of the class and teaching them a little chemistry or geology or whatever. You don't need a PhD in chemistry to talk about molecules or a PhD in geology to talk about soil, rock, and water. But it might feel funny at first because it's so different from what we mathematicians usually do. Almost all of the science in this book, especially in Part 1, is high school or college freshman level. Now there's nothing wrong with having a guest speaker on these topics, but there's also something very valuable, in my opinion, in having the students hear an actual math professor talking about these other topics.

Current events

Try to read the newspaper or new magazines with a view to what's going on currently in the environment. You may even want the students to do the same. I find that without any coaxing, they will sometimes come to class and mention that they heard of some problem that relates to the material being done in class, like an exciting "overturning of a gasoline tanker" during the morning rush hour. (One tends to develop a morbid interest in certain kinds of accidents while working on the material of Chapters 4 and 7.) You can even subscribe to current environmental news on on-line computer networks. These kinds of discussions in class bring a real currency to the course.

Testing and grading

Teachers usually have their own individual approaches to this. I give a weekly problem set, sometimes designated as an individual set (although they can still talk to each other) and sometimes a group set (only one paper from the group). I usually give an in-class final, divided into two parts, one closed book involving short answers, and the other on which book and notes and computer can be used. I also frequently offer extra credit problems or projects.

Computers

You could cover a lot of this material without using the computer, but that might bypass an opportunity for the students to gain experience in using this vital tool. There is a natural progression in the three themes treated in Chapters 2 through 4.

Chapter 2 (ground water) can easily be done without computers, but they are useful both for some of the repetitive calculations and for on-line research of certain environmental questions. Having computers in class is not essential, but it can be helpful for some teacher demonstrations and for student practice or group problem solving. This depends, of course, on availability.

Chapter 3 (air pollution) is aided significantly by the availability of programmable calculators or computers, as there are many repetitive calculations involving somewhat complicated exponential expressions. A spreadsheet program is available for users of the text, or the students can develop their own.

Chapter 4 (hazardous materials) requires computer access to run one of several commercial or public domain programs involving hazardous materials. One such program is available for users of the text, but it may be desirable to adopt a program actually in use in one's own community. (This has led to internship opportunities for some of our students.) One could skip the actual computations, but even in an advanced course where the emphasis is on the related Chapter 7, I think this would be a mistake.

Later chapters also use the computer as a tool for certain problems.

The instructions for accessing the air pollution spreadsheet and the hazardous material program may vary with time. Information is provided at the time of book purchase or can be obtained from the author.

The really nice thing about the use of computers in this kind of course is that it is not at all artificial, as it sometimes tends to be in elementary courses. The students really want the computer to help them solve the problems.

Group work

I have found it quite challenging to use group work in the classroom in many other math courses, either because there was a tight syllabus I felt responsible to cover (because the next course would be based on it), or because the students were not really motivated to work together. While these difficulties can be overcome, this particular material lends itself to group work. In particular, there are quite a few examples or problems that are introduced with a question, such as: "if x increases, would you expect y to increase, decrease, or stay the same." This is a good group question because then you can have each group defend its position in turn at the blackboard. I even have my groups pick team names to foster the development of some *esprit de corps*. (They laugh at first, but they get into it as we proceed.)

Mathematical gems for more advanced readers or courses

There are a few mathematical ideas in Part 2 of the text that I think of as real gems and that should perhaps be called to the attention of students, teachers, or other readers of this material. They have the capacity to expand one's mathematical imagination. They are treated partly in the main text and partly in this volume (through the solution of exercises), but I decided to highlight them here so that they are especially called to the attention of teachers. These include:

- The introduction and investigation of two of the most fundamental partial differential equations, Laplace's equation and the diffusion equation, from the point of view of current environmental issues and without any assumed prior background in differential equations. I think that for many students the focus on these equations and their investigation from multiple points of view (physical, probabilistic, continuum modeling) is more valuable intellectually than a rapid survey of a wider range of equations and boundary conditions. In Chapter 7, a heat flow problem arises for which the solution builds on the earlier analogous work for diffusion.
- The overlap or interaction between probabilistic models and continuum models, in this case in the use of elementary probabilistic methods to derive heuristically, but in fairly complete logical detail, the diffusion equation for a point source problem. (Along the way, several basic but very important statistical concepts are touched upon — the law of large numbers, the central limit theorem, and the binomial approximation to the normal distribution.) This is a common theme in more advanced courses, where the solutions to partial differential equations are often viewed as limiting cases of stochastic processes, but students too rarely see it in elementary courses.
- The solution of an important case of the one-dimensional diffusion partial differential equation from first principles so that no "tricks" from a prior partial differential equations course need to be called upon; much can be learned by wrestling with this problem on one's own even if, in the end, one might need a little help to put it all together. This is probably the hardest exercise in the text, but I have had several students complete it correctly (and each always somewhat differently).
- The use of iterative methods in a fixed point framework to solve complex equations, whether in the form of high dimensional linear systems (with a specific case arising from a ground-water problem) or one-dimensional nonlinear equations. Newton's method is reinterpreted this way.

- The clever idea of using "virtual sources" and symmetry concepts to deal with more complex boundary value problems.
- Various logical approaches to creating equations or functions to model a set of spatial data.

The range of these important concepts could perhaps make this text, emphasizing the second part, a novel alternative basis for an upper division introduction to applied mathematics.

Advice

I stand ready to respond to any inquiries about teaching this material, and I would be delighted to receive copies of additional related teaching materials and problem sets that I might make available to others. My current E-mail address (chadlock@bentley.edu at the time of this writing) and phone number should be easy to find, and, in any event, could always be obtained through the MAA.

Contents

1

Introduction*

I devote one class to the material in Chapter 1 of Volume 1, as I think it really starts to draw the students into the important environmental issues of the day.

One suggestion for how to introduce this material in an interesting way would be to ask the students to make a list of all the environmental issues they can think of, and then, perhaps after 10 minutes, take suggestions from the floor and start to compile a master list on the blackboard. You might even use categories: global, national, local, etc.

Discuss each issue as it comes up, asking other students to add things before you take over. If you don't know much about an issue, don't be afraid to say so; but it will be very helpful if you can really teach them some interesting things about these issues right from the start.

Things you should be sure to get straight in your mind are the greenhouse gas effect and the ozone issue, which are different, and any other issues you are sure will come up (e.g., rain forests). A little extra reading on the teacher's part here will go a long way. A very good type of source for you to have available for such preparation would be an environmental science text, where you will usually find capsule summaries of each issue, along with good diagrams.

After this list is compiled, you could go back through it in order, asking them if they can think how mathematical models might be used for each of the issues. You can probably base this on your common sense, combined with the information in Table 1-1 in Volume 1.

(No exercises.)

---

* Note: Vertical lines at the margins are used throughout to denote general comments and suggestions that are not part of actual solutions to individual exercises.

2

Ground Water

2.1 Background

It is important to keep in mind that the objective of this book and most courses using it is to teach both about environmental issues *and* mathematics. Don't feel that you have to rush through the environmental background just so you can do some mathematics. Asking students to do some of the research questions contained below and elsewhere gets them really immersed in these issues. In fact, the open-ended nature of some of the questions causes them to read news articles, seek out information sources, and ask others questions. It can sometimes be frustrating for them as they do this, since they are so used to math problems that have very specific answers, but they rapidly come to develop an involvement in these issues. Thus they approach the later math modeling with much better motivation. Take your time discussing these issues in class and try to serve as a cheerleader and resource as you send them off to research the issues. You may need to help them with their computer search skills or invite a speaker to class from the library to do a demonstration.

You may also find that setting up groups within the class works well both on the research problems and on many of the numerical problems, the main drawback being that students with weak research skills may not ever really do any of the research themselves. Thus you might want to have some individual and some group assignments.

For research questions such as those below, one way to proceed is to use an on-line computer search program, such as are available in many libraries and on many computer networks. Another approach is to look through the telephone directory for relevant government agencies or other possible sources of information. Library reference departments are also excellent sources of help.

1. **Identify at least one government-recognized site with contaminated ground water in either your home town or a town that borders it, or your home country (if not the US). In one page, summarize its status in terms of being investigated and/or cleaned up. You should identify the contaminants of interest there, how the contamination arose, and other items of interest. Almost every community has such sites.**

Some people find an interesting site somewhere by doing a computer search, and then they write it up, failing to adhere to the instructions about its being their home town or adjoining the latter. The intention of the question as written is to make them dig a little deeper so as to see that such sites can be found almost everywhere. Furthermore, the question asks for contaminated ground water, not just contaminated soil, such as might result from a spill or leak that was caught soon after it occurred. Past students have been quite resourceful on this. For example, they have telephoned or sent E-mails to parents or friends at home, if far away, who then have made inquiries and sent back relevant data, even for out-of-the-way locations in Central America and Cyprus. In some cases this has led to interesting conversations between students and their parents or neighbors. In fact, computer searches often yield such sites, at least in the US, with minimal effort; but it is also very motivating to have established a more personal communication with people connected to a site. Class discussions of the results of this exercise can be quite interesting, both on the substance itself and on the various search methods that succeeded or failed along the way. Federal Superfund sites can be found by accessing the EPA's web page or equivalent (www.epa.gov at the time of this writing, but you can always use an online search tool so you don't have to remember such addresses or check for changes or updates). You will find a one- or two-page fact sheet on every site on the National Priorities List, as well as much additional site information. Not every community has Federal Superfund sites, but there are even more state-recognized sites in much of the country. On-line newspaper indexes are a great way to find these, using search terms such as: [town name] AND contamination. Naturally, for classes wishing to be less computer dependent, there are excellent resources available through the library (which lets the students see how helpful reference librarians can be) and the phone book.

2. Compile a list of all government-recognized contaminated underground sites in your home city or town. (You will probably be quite surprised at this number.) Not all of these sites necessarily have contaminated ground water, because many of them are sites where there have been leaky underground storage tanks which have since been removed and where some contamination was found in the soil around the tanks. It is not always the case that this contamination has moved downward to the water table where it actually enters the ground-water system.

Use the resources described in the previous problem and at the beginning of the exercise set. These lists often include sites where some past event occurred that might have led to contamination, but where no actual investigation or only an incomplete investigation has taken place to date. That's why you can't tell whether there is actual ground-water contamination. A town or city with a history of past industrial activity is almost assured of having a number of such sites, but, in fact, almost any town with gas stations has such sites. It's generally harder to get state lists, and sometimes you have to visit the state agency responsible for environmental affairs to do so. However, such lists have occasionally been published in the newspaper, hence providing an alternative source. Environmental professionals in the local area (see the yellow pages under "Environmental") might quickly direct one to good sources, or even provide the information.

An alternative way to present this question is to apply it to the town or city containing your school, so that the teacher can do some advance work and decide whether to ask for all sites, just federal sites, etc. If the EPA has a local office, the librarian there is often very accustomed to helping students find information of this type as well.

3. Compile a list of all National Priority List (NPL) sites identified by the EPA within your home state, and for three such sites provide a brief summary of the nature of the contamination.

The advantage of this over the previous problems is that it can be solved very conveniently by on-line access to the EPA's resources. However, it does have the disadvantage that it does not necessarily engage the students in a site in their own home town or demonstrate how widespread such contaminated sites are. Thus the problems are really complementary.

As an example of the information available, here is the site list for Massachusetts as of this writing:

Site Name: ATLAS TACK CORP.
Location : Bristol County Fairhaven
EPA ID# : MAD001026319

Site Name: BAIRD & MCGUIRE
Location : Norfolk County South Street in northwest Holbrook
EPA ID# : MAD001041987

Site Name: NAVAL WEAPONS INDUSTRIAL RESERVE PLANT
Location : Middlesex County Bedford
EPA ID# : MA6170023570

Site Name: NEW BEDFORD SITE
Location : Bristol County 55 miles south of Boston
EPA ID# : MAD980731335

Site Name: BLACKBURN AND UNION PRIVILEGES
Location : Norfolk County Walpole
EPA ID# : MAD982191363

Site Name: CANNON ENGINEERING CORPORATION
Location : Plymouth County Bridgewater
EPA ID# : MAD079510780

Site Name: CHARLES-GEORGE RECLAMATION TRUST LANDFILL
Location : Middlesex County 30 miles northwest of Boston
EPA ID# : MAD003809266

Site Name: FORT DEVENS
Location : Worcester County & Middlesex County 35 miles west of Boston
EPA ID# : MA7210025154

Site Name: FORT DEVENS-SUDBURY TRAINING ANNEX
Location : Middlesex County Portions of the Towns of Sudbury, Maynard, Hudson, and Stow
EPA ID# : MAD980520670

Site Name: GROVELAND WELLS
Location : Essex County Groveland
EPA ID# : MAD980732317

Site Name: HANSCOM FIELD/HANSCOM AIR FORCE BASE
Location : Middlesex County Towns of Bedford, Concord, Lexington, and Lincoln
EPA ID# : MA8570024424

Site Name: HAVERHILL MUNICIPAL LANDFILL
Location : Essex County 2 miles southeast of downtown Haverhill
EPA ID# : MAD980523336

Site Name: HOCOMONCO POND
Location : Worcester County Westborough
EPA ID# : MAD980732341

Site Name: INDUSTRI-PLEX
Location : Middlesex County North Woburn
EPA ID# : MAD076580950

Site Name: IRON HORSE PARK
Location : Middlesex County North Billerica
EPA ID# : MAD051787323

Site Name: NORWOOD PCBs
Location : Norfolk County Kerry Place in Norwood
EPA ID# : MAD980670566

Site Name: NYANZA CHEMICAL WASTE DUMP
Location : Middlesex County Megunco Road in Ashland
EPA ID# : MAD990685422

Site Name: OTIS AIR NATIONAL GUARD/CAMP EDWARDS
Location : Barnstable County Falmouth
EPA ID# : MA2570024487

Site Name: PLYMOUTH HARBOR/CANNON ENGINEERING CORP.
Location : Plymouth County 11/2 miles northwest of Plymouth Center
EPA ID# : MAD980525232

Site Name: PSC RESOURCES
Location : Hampden County Palmer
EPA ID# : MAD980731483

Site Name: RE-SOLVE, INC.
Location : Bristol County North Dartmouth
EPA ID# : MAD980520621

Site Name: ROSE DISPOSAL PIT
Location : Berkshire County Lanesborough
EPA ID# : MAD980524169

Site Name: SALEM ACRES
Location : Essex County Salem
EPA ID# : MAD980525240

Site Name: SHPACK LANDFILL
Location : Bristol County On the Attleboro/ Norton town line
EPA ID# : MAD980503973

Site Name: SILRESIM CHEMICAL CORP.
Location : Middlesex County Lowell
EPA ID# : MAD000192393

Site Name: SOUTH WEYMOUTH NAVAL AIR STATION
Location : Norfolk and Plymouth Counties Towns of Weymouth, Abington and Rockland
EPA ID# : MA2170022022

Site Name: SULLIVAN'S LEDGE
Location : Bristol County New Bedford
EPA ID# : MAD980731343

Site Name: MATERIALS TECHNOLOGY
LABORATORY (USARMY)
Location : Middlesex County Watertown
EPA ID# : MA0213820939

Site Name: NATICK LABORATORY ARMY
RESEARCH, DEVELOPMENT, AND
ENGINEERING CENTER
Location : Middlesex County Natick
EPA ID# : MA1210020631

Site Name: W. R. GRACE & CO., INC. (ACTON
PLANT)
Location : Middlesex County Off Independence Road
in Acton and Concord
EPA ID# : MAD001002252

Site Name: WELLS G & H
Location : Middlesex County City of Woburn
EPA ID# : MAD980732168

A typical site summary from the same database is the following, corresponding to the first site in the above list:

ATLAS TACK CORP.
    MASSACHUSETTS
    EPA ID# MAD001026319
    EPA REGION 1
    Bristol County
    Fairhaven

*** This fact sheet reflects site information as of: January 1997 ***

Site Description

The Atlas Tack Corporation formerly manufactured cut and wire tacks, steel nails, and similar items on a 12-acre site in Fairhaven. From the 1940s until the late 1970s, wastes containing cyanide and heavy metals, including high levels of arsenic, were discharged into an unlined acid neutralizing lagoon located approximately 200 feet east of the manufacturing building and adjacent to a saltwater tidal marsh in Buzzards Bay Estuary. Other contaminated areas at the site include a filled wetland, former dump, and other chemical spills. The area is residential and commercial. Approximately 7,200 people live within a 1 mile radius, and approximately 15,150 people live within 3 miles of the site.

Site Responsibility:

The site is being addressed through Federal and potentially responsible parties' actions.

NPL LISTING HISTORY

Proposed Date: 06/24/88
Final Date: 02/21/90

Threats and Contaminants

The ground water is contaminated with cyanide and toluene that leached from the site lagoons. The on-site soil is contaminated with volatile organic compounds (VOCs), including toluene and ethyl benzene; heavy metals, including chromium, cadmium, lead, and nickel; pesticides; polychlorinated biphenyls (PCBs); and polycyclic aromatic hydrocarbons PAHs). Trespassers are at risk through direct contact with contaminated soil or ingestion of shellfish in the area. The marsh south of the lagoon and estuarine areas in Buzzards Bay are also contaminated causing an ecologic risk to the wildlife.

Cleanup Approach

The site is being addressed in two phases: initial actions and a long-term remedial phase focusing on cleanup of the entire site.

Response Action Status

Initial Actions: In late 1992, the potentially responsible party installed a fence around the site to control access.

Entire Site: The EPA is conducting an investigation into the nature and extent of site contamination. Currently, the EPA is conducting the feasibility study to identify appropriate alternatives for cleaning up the site, with a final remedy expected to be selected in mid-1998.

Environmental Progress

The EPA has determined that the public and the environment are not at immediate risk while investigations at the Atlas Tack Corp site continue and final cleanup alternatives are being determined.

Site Repository

Fairhaven Public Library, Center Street, Fairhaven, MA 02719

TABLE 2-A
Summary of progress on NPL sites

Stage of activity	Number of sites	Percentage of total
Clean-up remedy construction complete	346	25%
Construction of clean-up remedy underway	472	34%
Design of clean-up remedy underway	169	12%
Clean-up remedy selected	82	6%
Site investigation or emergency clean-up action underway	303	22%
Total number of NPL sites	1372*	

* The total is less than sum of the previous numbers because some sites are treated as multiple units, which may be in different stages of remediation.

4. Find out how many sites on the National Priority List (NPL) published by the EPA have actually been completely cleaned up to the government's satisfaction or have at least completed the process of construction of facilities for long term remedial activity. For one NPL site in your home state (every state has at least one), determine the actual or projected cost of cleanup.

This changes with time, of course, but access to the EPA web page at the time of this writing (1997) leads to the summary information shown in Table 2-A.

Most complex site remediations, such as on NPL sites, involve long-term activities on site, such as ground-water "pump and treat" plants which may be expected to pump out ground water for twenty years or more, clean it by various methods, and pump it back into the ground or into surface water, or incineration plants through which contaminated soils or sediments from the site will gradually be processed to destroy their toxic contents.

Costs vary widely, but can be very high. One generally wants to access the ROD (Record of Decision) for the site, which is the formal document representing the agreed upon action to be taken. (The RODs are all available on-line via the EPA Superfund resources). For example, the "Wells G and H" site in Woburn, Massachusetts, the subject of a Nova television program and a number of books, is divided for remedial purposes into three "operable units." The estimated cost of remediation of the first unit alone is $70 million, according to the ROD, including both capital costs and 30 years of operating costs converted to present values. The other units have not yet been finalized. (The author's experience is that the ROD estimates are often exceeded by a substantial factor because further complications tend to develop once underground excavations or other actions begin.)

2.2 Physical Principles

For the exercises below that pertain to your own community, good local sources of information are your municipality's public works department, local science teachers, civil engineers that work in the area, or environmental agencies or organizations. Environmental science texts and environmental reference books are good sources for more general information.

1. For your own home residence, determine where the drinking water comes from. If it comes from ground water, determine the general nature of the aquifer formation being used.

In general in the US, about 50% of our drinking water comes from ground water, much more so in rural areas. (Facts like this are commonly located in environmental science texts. For example, this statistic is from p. 235 of G. Tyler Miller, Jr.: *Environmental Science*.) Such residential ground water may come from individual wells, either deep, drilled wells into rock or soil aquifers, perhaps up to 600 feet deep, or shallow "dug" wells less than 25 feet deep. If the residence is supplied by a municipality, it may utilize a combination of surface and ground-water sources, and its wells would need to tap high productivity aquifers, such as sands and gravels along a river valley, porous limestone formations, or similar formations.

2.   Determine the general allocation of water supplies in the US to various distinct purposes. (Consumption is only a small fraction.)

From the Miller book cited above, p. 227, uses are divided into public (10%), industry (11%), electric cooling (38%), and agriculture (41%). There are various ways to account for things, of course. For example, electric cooling is primarily the use of water to cool the condensers of steam generating plants, from where it is returned to circulation at a slightly elevated temperature. Irrigation is of great concern, because in the plains states the Ogallala Aquifer is being "mined" for irrigation water, meaning that water that has been stored there since the last glacial period is now being depleted in an irreversible way that will eventually catch up with us. Thus some hydrologists distinguish between consumptive and nonconsumptive uses. About every five years, the US Geological Survey publishes a report on water use in the US. Approximately a fifth of all freshwater use in the US is supplied from ground-water sources.

3.   Determine the extent to which ground water is used in your own home community for various purposes. To what extent is it treated or tested before domestic use?

The uses will vary greatly depending on the part of the country under discussion; see the previous response for various possibilities.

With regard to testing and treatment, if the water is from an individual well, it is usually not treated or tested. In some areas of the country some treatment might be provided for iron removal or softening or other conditioning purposes. Some people have their water tested periodically, especially if they have concerns. Depending on possible contamination sources, it might be tested for fecal coliform bacteria, organics, or metals. Municipal well water is more likely to be tested regularly and to be treated with agents like chlorine to retard microbial growth during storage and distribution.

2.3 Typical Quantitative Issues

(No exercises.)

2.4 Darcy's Law

1.   Imagine an underground sand aquifer that is 30 feet thick, 400 feet wide, and has a hydraulic conductivity of 30 ft/day. Two test wells 550 feet apart have been drilled into the aquifer along the axis of flow, and the measured head values at these wells were found to be 115 and 103 feet respectively.

   a)   Draw a really clear diagram of this situation.

A good model for a diagram is the block diagram in Figure 2-15, which captures the full three dimensional nature of the problem. One would need to change only the numbers.

   b)   Find the total flow rate in a given cross section of this aquifer.

$$Q = K \times i \times A$$
$$= 30 \text{ ft/day} \times \frac{115 \text{ ft} - 103 \text{ ft}}{550 \text{ ft}} \times (30 \text{ ft} \times 400 \text{ ft})$$
$$= 30 \text{ ft/day} \times .02182 \times 1200 \text{ ft}^2$$
$$= 7855 \text{ ft}^3/\text{day}$$

2. **This problem refers to the situation of Figure 2-14. In addition to the information provided in that figure, suppose that the aquifer is composed of a coarse porous sand. What would be the volumetric flow rate through any single imaginary planar surface of area one square foot, oriented perpendicular to the direction of flow? Describe how such a surface would be oriented with respect to the plane represented by Figure 2-14.**

Such an area would be oriented perpendicular to the flow arrow shown on the figure, so that one of its axes would be coming out perpendicular to the paper. It could have any shape: square, rectangular, round, or irregular. We use the high end of the sand hydraulic conductivity range from Table 2-1, namely 1000 ft/day. The flow rate would be:

$$Q = K \times i \times A$$
$$= 1000 \text{ ft/day} \times \frac{(100 - 16) \text{ ft} - (96 - 18) \text{ ft}}{350 \text{ ft}} \times 1 \text{ ft}^2$$
$$= 1000 \text{ ft/day} \times .0171 \times 1 \text{ ft}^2$$
$$= 17.1 \text{ ft}^3/\text{day}.$$

Sometimes a student will use a different value of K, perhaps on the high side but not at the extreme. This would be perfectly acceptable as long as there is some rationale for the choice of a value. A coarse porous sand could still vary in K value, depending on the particle shapes and the presence of fines (smaller grains).

This problem also introduces a unit issue to be elaborated on later in the text. The flow rate is in units of ft^3/day, but this is *per square foot* of cross-sectional area. Canceling out the units of ft^2 would yield a value of 17 ft/day, which looks like a velocity! It is not, but it is a common source of confusion. This will be discussed later in the text; this is just an advance notice.

3. **Consider the aquifer shown in Figure 2-15 and for which a volumetric flow calculation was carried out in the text. If an engineer calculates that over the past five years, an old landfill above the aquifer has caused about 100 pounds per year of a certain contaminant to leach into the aquifer, what would this imply for the value of the resulting long-term average concentration in the aquifer? Express your answer in the following three sets of units: lb/ft^3, g/cm^3, and ppm.**

If the contamination rate is 100 lb/yr, this can be converted to a daily rate of 100/365 lb/day. Following the approach illustrated in the text, we can thus obtain the concentration as follows:

$$C = \frac{\left(\frac{100}{365}\right) \text{ lb/day}}{7,333 \text{ ft}^3/\text{day}} = \frac{0.274 \text{ lb/day}}{7,333 \text{ ft}^3/\text{day}} = 3.7 \times 10^{-5} \text{ lb/ft}^3.$$

Beginning with this value, we convert to the other two systems of units as follows:

$$\frac{3.7 \times 10^{-5} \text{ lb}}{\text{ft}^3} \times \frac{453.6 \text{ g}}{1 \text{ lb}} \times \left(\frac{1 \text{ ft}}{12 \text{ in}}\right)^3 \times \left(\frac{1 \text{ in}}{2.54 \text{ cm}}\right)^3 = 5.93 \times 10^{-7} \text{ g/cm}^3$$

$$\frac{3.7 \times 10^{-5} \text{ lb (con.)}}{\text{ft}^3 \text{ (water)}} \times \frac{1 \text{ft}^3 \text{ (water)}}{62.4 \text{ lb (water)}} \times \frac{10^6 \text{ lb (water)}}{1 \text{ million lb (water)}} = \frac{0.6 \text{ lb (con.)}}{\text{million lb (water)}} = 0.6 \text{ ppm}.$$

4. **Consider the aquifer discussed in Exercise 1, above. Suppose it is discovered that a chemical pipeline passing over the aquifer has apparently had a small leak for a long period of time. Furthermore, by comparing chemical inventory records at both ends of the pipeline, it is determined that about 10 pounds per week have probably been lost through this leak. What would be the resulting long-term average concentration of the chemical in the aquifer? Express your answer in the following three sets of units: lb/ft^3, ppm, and kg/m^3. (Note: "kg" refers to kilograms, or one thousand grams; "m" refers to meters.)**

In this case, we also begin by converting the contamination rate to a daily rate of 10/7 lb/day. Then we obtain the concentration as follows:

$$C = \frac{\left(\frac{10}{7}\right) \text{ lb/day}}{7,855 \text{ ft}^3/\text{day}} = \frac{1.43 \text{ lb/day}}{7,855 \text{ ft}^3/\text{day}} = 1.8 \times 10^{-4} \text{ lb/ft}^3.$$

Beginning with this value, we convert to the other two systems of units as follows:

$$\frac{1.8 \times 10^{-4} \text{ lb (con.)}}{\text{ft}^3 \text{ (water)}} \times \frac{1 \text{ ft}^3 \text{ (water)}}{62.4 \text{ lb (water)}} \times \frac{10^6 \text{ lb (water)}}{1 \text{ million lb (water)}} = \frac{2.9 \text{ lb (con.)}}{\text{million lb (water)}} = 2.9 \text{ ppm}$$

$$\frac{1.8 \times 10^{-4} \text{ lb}}{\text{ft}^3} \times \frac{453.6 \text{ g}}{1 \text{ lb}} \times \frac{1 \text{ kg}}{1000 \text{ g}} \times \left(\frac{1 \text{ ft}}{12 \text{ in}}\right)^3 \times \left(\frac{1 \text{ in}}{2.54 \text{ cm}}\right)^3 \times \left(\frac{100 \text{ cm}}{1 \text{ m}}\right)^3 = 2.9 \times 10^{-3} \text{ kg/m}^3$$

2.5 Interstitial Velocity Equation

The issue of units will become more and more important as we move on, and one question is whether to provide all necessary unit conversions to the students or to let them develop some resourcefulness in finding their own. Even the number of feet in a mile, which shows up below, may not be known; and it is very likely unfamiliar to foreign students. The dictionary is a great resource. One doesn't need a whole table of values, only individual conversions that let you get from your starting place to where you need to be. Some spreadsheet programs, such as Excel, have a built in conversion function. Some students have responded well to the suggestion that they construct their own unit conversion spreadsheet program, adding new conversions to it as they encounter the need throughout the course. This "personalization" of the program seems to facilitate their achieving greater mastery of the whole process.

1.   A homeowner notes a strange taste entering into his well water, which comes from a deep well in a bedrock aquifer over 250 feet deep. The nearest source of contamination he can think of is a gasoline station three-quarters of a mile away. He contacts an attorney, who in turn hires an engineering consultant to collect and analyze available data relevant to this situation. The data are as follows. The house and the gasoline station are underlain by a bedrock aquifer, and the hydraulic head according to published ground-water contour maps for the area is 38 feet at the house and 44 feet at the service station. The hydraulic conductivity of the aquifer has been estimated in a regional study to be about 3 ft/day, and the porosity (corresponding to the fracture system) to be about 0.001. The tanks at the gasoline station are pressure-tested once a year for leakage, and the last test, which still showed no leakage, took place 7 months before the contamination was noticed by the homeowner. What do you think the consultant's analysis showed? Should the homeowner pursue his case?

First, notice that there is no clear indication that the ground water is flowing in a line from the gas station to the well; it could well be going at some oblique angle to the line connecting them. Nevertheless, the first question is whether the gas station *could possibly* be the source, so we will test this under the assumption that flow is indeed in this direction. The velocity and travel time are found to be:

$$v = \frac{Ki}{\eta} = 3 \times \frac{44 - 38}{.75 \times 5280} \times \frac{1}{0.001} = 4.55 \text{ ft/day}$$

$$t = \frac{3960}{4.55} = 870 \text{ days} = 2.4 \text{ years to cover the full distance.}$$

Some students alternatively use the seven-month time period and calculate that in that time the material could cover only about 970 feet, less than the distance to the well. Note that the depth of 250′ is not relevant to the above calculation, but its appearance in the problem will cause them to think about whether it is relevant.

This is a good problem for discussion, which the last part of the question can be used to lead into. In particular, there are very many uncertainties in the analysis. The head and conductivity values are from regional studies, and the latter can vary widely over several orders of magnitude, as shown in Table 2-1. It would take only a factor of 4 increase in K to yield a travel time of seven months for the full distance, and a factor of 10 would even allow the leak to occur much later than just after the last tank pressure test. On the other hand, the determination of site specific values could be quite expensive, especially given the depth of the aquifer. The homeowner might be able to have further chemical analyses done on the contaminated water, with the hopes of picking up an identifiable chemical signature, but it would be practically impossible to assign it to this one gas station. It could be even much older contamination finally reaching the well from some earlier unrecognized source. The homeowner could collect some additional data by testing the water from any houses in the vicinity of his own, where some pattern of concentration levels might be seen. He could also see if there might be any other wells in the area for him to use to determine the local head distribution, but this could be a complex process.

It is likely that the gasoline station would want to or would be willing to undergo a new tank pressure test. If the results are negative (no leak), it is still possible that the contamination might have come from an earlier leak of older tanks and residual soil contamination from that period. The history of this could be investigated. If the tank tests show a new leak, then in the process of remediating the problem, the soil around the tank would be tested to see how far the contamination might have spread. Since fractured rock is very unpredictable in terms of fluid transport (e.g., one long fracture might serve as a conduit for transport), the service station owner or his insurance company might be willing to negotiate a settlement with the homeowner if indeed it looks like: a) some gasoline-like contaminant got to his house, and b) some did seem to escape from the immediate locale of the tank. (This problem is based on a real case that the author was involved in.)

Thus in this case the hydrologic analysis is not definitive, but it does indicate whether one is in the "ball park" for a possible connection.

2.   Referring to the situation of Figure 2-14 and its further elaboration in Exercise 2 of Section 2-4, calculate the actual ground-water velocity.

For coarse, porous sand we use the high extreme of porosity from Table 2-1, namely 0.5. Thus, using the same K as in the earlier solution, the velocity would be:

$$v = \frac{K \times i}{\eta} = \frac{1000 \text{ ft/day} \times .0171}{.5} = \frac{17.1 \text{ ft/day}}{.5} = 34.2 \text{ ft/day}.$$

3.   Suppose you were trying to pin down further the ground-water travel time in Exercise 1, above, and you had a fixed amount of money available to invest in technical services to refine your estimates of the hydrologic parameters for the locality. In general, do you think it would be more productive to direct the money towards improving your estimate of hydraulic conductivity K or of porosity η, if this were the choice you were given? (Assume that the money would be equally effective in narrowing the uncertainly band for each of these parameters.)

In terms of the interstitial velocity equation, the results vary similarly with respect to K and η. For example, a factor of two increase in K has the same effect as a factor of two decrease in η. However, it is clear from Table 2-1 that there is a much wider range of values for K than for η for almost any kind of geologic formation. Therefore, if all you know is the kind of formation, you generally have more uncertainly in the K value, and therefore this would be a more reasonable target for your further stage of data collection. As a practical matter, for fractured bedrock aquifers, as in Exercise 1, the use of Darcy's law is generally better over longer and longer distances, where the "average" fracture characteristics tend to control the flow, rather than one dominant fracture, which can serve essentially as a conduit over relative short distances.

These kinds of issues are faced all the time in real site investigations. A related question is how best to refine the parameter values, where you might need to make a choice between getting good "very local" values by drilling individual wells and studying the formations with TV cameras, logging devices, or hydraulic tests, and sometimes more effective, "average" area values obtained by seeing what modeling value is most consistent with other measurable parameters, such as head values in given wells. This latter approach is pursued further in Chapter 5.

4.   Discuss the following hypothesis: in general, lower porosity geologic media usually tend to have lower ground-water velocities.

This is a good problem for discussion because you can make an advance argument for either the hypothesis or its opposite. For example, lower porosity should tend to obstruct flow and hence slow it down. On the other hand, a lower porosity in the interstitial velocity equation's denominator would tend to increase v!

TABLE 2-B

Calculation of the quotient K/η at porosity/conductivity extremes

Geologic medium	Low end (low porosity and low conductivity)	High end (high porosity and high conductivity)
Gravel	400	250,000
Sand	0.04	2
Silt	3×10^{-4}	0.2
Clay	2.5×10^{-7}	1.4×10^{-3}
Sandstone	2×10^{-4}	0.33
Limestone	0.001	0.01
Granite	1	100

From the interstitial velocity equation, it is clear that the velocity depends on the quotient K/η for a given hydraulic gradient i. Making the assumption that for *a given kind of geologic medium,* lower porosity often corresponds to lower hydraulic conductivity, we can calculate the quotient K/η using both the low and high end porosity/conductivity extremes from Table 2-1, the results of which are shown in Table 2-B.

This table suggests that certainly within a given kind of formation, lower porosity would tend to correspond to lower velocity. Although this sounds intuitively reasonable, it does seem to be at variance with the interstitial velocity equation, where η is in the denominator (so that smaller values would tend to increase velocity). The explanation, of course, is that lower porosity is correlated with lower conductivity, which more than compensates for the porosity reduction in the denominator. Even across formations, the correlation often holds, with most of the smaller values in the table coming from the low porosity ends of the extremes. However, note the case of granite, dominated by fracture flow, where even the low end value is practically as large as the high end value for sand! (But while the velocities may be comparable, the volumetric flow rates within sand would be much larger.)

Naturally, all of the above discussion is predicated on having comparable hydraulic gradients, not unreasonable since these latter depend more on topography and regional factors, which will often be similar even if the underground formations are different.

2.6 Discussion of Parameters

(No exercises)

2.7 Use of Head Contour Diagrams

One way to introduce this topic is to pass out a few copies of a topographic map and have the students work in groups to answer some questions one at a time. (E.g., are there any cliffs? Where is there a large flat plain? Which way would a ball roll if dropped off at a given point? Why?) This would start to introduce some key ideas. You can also compare this kind of map to a 3-D relief map, a copy of which you might also want to take to class if one is available.

Good work organization is key to taking the tedium out of some of the following exercises, and perhaps the simplest way to do them is to use a computer spreadsheet. Since this is a tool that will be even more useful in the next chapter, this would be a good time to introduce it and make sure everyone is up to speed. The calculations can be done by hand or with a programmable calculator or other computer program as well, so you can adapt to almost any working environment. But even if you, the teacher, may not be conversant with spreadsheets, it might be a good time to experiment a bit. They are extremely easy. In addition, they will easily let you check student calculations that are based on a slightly different flow path or interpretation of the scale. They will also let you construct new problems and solutions easily!

This is also a good way to get students to take pride in clear presentation of results. Some really like to "jazz up" the format of their spreadsheets so they are both clear and very inviting. If you pick one of these to distribute as a sample of a nice assignment, soon you will have most of the class getting more involved in effective presentation of results.

1.   Consider the situation described by Figure 2-24. The contour lines represent estimated contours of hydraulic head in a shallow aquifer composed chiefly of coarse sand. If a major gasoline spill onto the ground occurs at point X, as shown on the figure, and some of the gasoline seeps down to the water table, indicate on the diagram its likely migration path with the ground water, and estimate how long it might take for the first traces of dissolved gasoline to reach the boundary of the region shown on the figure. Pick representative values of hydrologic parameters you need from Table 2-1. If you need to make any additional assumptions, be sure to explain what you do. (Note: the bulk of the gasoline would not be expected to mix with or dissolve in the ground water; it would accumulate near the top of the aquifer and move downgradient according to some complex processes. However, a small amount will actually dissolve in the water and have a noticeable effect on taste, even at very low concentrations.)

The general steps in the solution are:
1. Draw the flow path as a smooth curve, crossing each contour line perpendicularly. (There will of course be some variation in how the students do this, but smoothness and perpendicularity are what you are looking for.)
2. Using the scale of the map, measure the lengths of each segment of this flow path. The key thing here is to do it segment by segment, not just total length. That's because the gradient is changing a bit along the path. (If

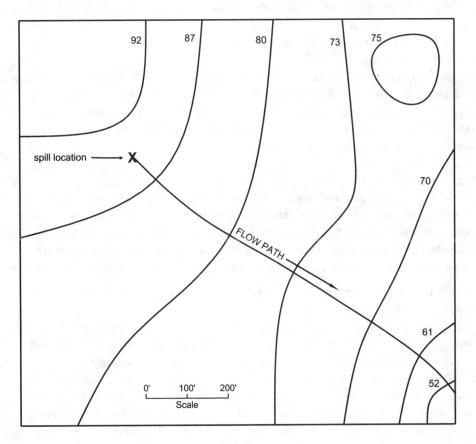

FIGURE 2-A
Flow path for Exercise 1

TABLE 2-C
Spreadsheet for Exercise 1

GROUND WATER TRAVEL TIME FOR EXERCISE 1, SECTION 2.7								
Conductivity =		1,000 ft/day						
Porosity =		0.5						
Segment	Length	h1	h2	i(calc)	i(extrap)	i	v (ft/day)	t(days)	t(years)
1	84		87		0.031	0.031	62.6	1.3	0.0037
2	224	87	80	0.031		0.031	62.6	3.6	0.0098
3	175	80	73	0.040		0.040	80.1	2.2	0.0060
4	222	73	70	0.014		0.014	27.0	8.2	0.0225
5	139	70	61	0.065		0.065	129.0	1.1	0.0030
6	97	61	52	0.092		0.092	184.7	0.5	0.0014
7	25	52			0.092	0.092	184.7	0.1	0.0004
Total								17.0	0.0467

the gradient were constant, as it is in some later problems, then you would get the same answer both ways, as discussed in the text.) If you are using a spreadsheet, you could even just input the ruler length of each segment and let the spreadsheet do the distance conversion based on the scale factor you deduce from the scale on the figure. Be sure to remind the students that you are trying to estimate the curvilinear distance, not the straight line distance from crossing to crossing.

3. Use the distances together with the head values at the contours to calculate a gradient value for each segment. For end segments, where you are missing the head value at one end, use the adjacent segment's gradient, as in the example in the text, or you might extrapolate a head value at the endpoints, and then use this to get a gradient.

4. Now you can calculate velocity and hence travel time for each segment.

5. Then you sum the travel times.

The calculations are based on the flow path shown in Figure 2-A, and the final result is a travel time of about 17 days, derived from the calculations shown in Table 2-C. (In the real world, this would be interpreted as "in a few days or weeks, probably less than a month, and certainly well less than a year," given the uncertainties associated with the various parameters, which can vary within a given formation.) Note that students might have used slightly different parameter values from Table 2-1, which is fine.

2.   **Consider the situation described by Figure 2-25. The contour lines represent estimated contours of hydraulic head in a shallow aquifer composed chiefly of a slightly silty medium sand. If a leak of hazardous plating wastes (which may contain heavy metals such as chromium and cadmium) occurs and seeps into the ground at point X, as shown on the figure, indicate on the diagram the likely migration path with the ground water, and estimate how long it might take for the first traces of contamination to reach the boundary of the region shown on the figure. Use reference values from Table 2-1 to choose your parameters.**

The flow path is sketched in Figure 2-B, and the calculations are shown in Table 2-D. Based on the input parameters chosen, the answer would be about 186 years.

Note that in this case, even more so than in the previous problems, students might choose quite varied input parameters. Since the objective of the course is not to make them professional hydrologists, one does not have to overemphasize parameter selection, as long as they have a rationale that is reasonable. In this case, the parameters shown in the table are based on the following considerations. Slightly silty sand is sand containing some silt, not a great amount. So it doesn't necessarily bring about a drastic reduction in conductivity, as if there were enough to plug up most of the pore spaces. The conductivity in this case depends as much or more on the size of the sand particles, which in this case is what the "medium" refers to. To allow room for fine sands, as well as sands with more substantial silt or clay content, on our scale, we would probably still be in the upper half (considered logarithmically on the scale of factors of 10), and so a value of 10 was chosen. Since the silt fills some of the pores

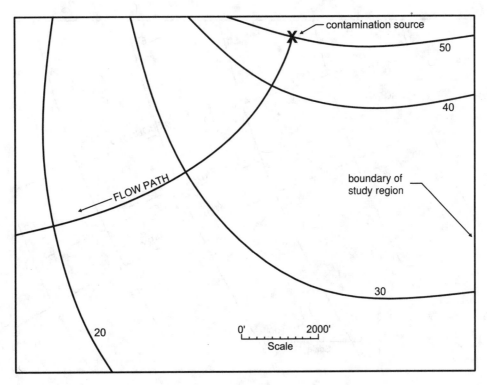

FIGURE 2-B
Flow path for Exercise 2

TABLE 2-D
Spreadsheet for Exercise 2

GROUND WATER TRAVEL TIME FOR EXERCISE 2, SECTION 2.7									
Conductivity =		10 ft/day							
Porosity =		0.3							
Segment	Length	h1	h2	i(calc)	i(extrap)	i	v (ft/day)	t(days)	t(years)
1	1,161	50	40	0.009		0.009	0.2870	4,046	11
2	2,850	40	30	0.004		0.004	0.1169	24,375	67
3	3,239	30	20	0.003		0.003	0.1029	31,467	86
4	840	20			0.003	0.003	0.1029	8,165	22
Total								68,052	186

between larger sand grains, a value on the low side for porosity was chosen, although not the actual extreme, since there could be lower porosity cases. This depends also on the shape of the grains and how well they pack together.

3. The diagram in Figure 2-26 shows the hydraulic head contours in the vicinity of a waste rock lagoon or "slimes pond" at a secondary gold mining facility, that is, a facility where old waste rock from an earlier gold mine is being reprocessed to pull out more of the gold. Suppose that the aquifer under the site averages 20 feet thick, its hydraulic conductivity is 190 ft/day, and its porosity is 0.25. Cyanide solution (used to separate gold) is known to be seeping from the lagoon into this aquifer. By considering rates of precipitation, evaporation, and slimes disposal, a consulting engineer has estimated that this seepage is occurring at a rate of roughly 50 cubic feet per day uniformly under the lagoon site, and this solution actually contains about 0.01% cyanide by weight. You are interested in both travel time to and concentration at a planned drinking water well at W.

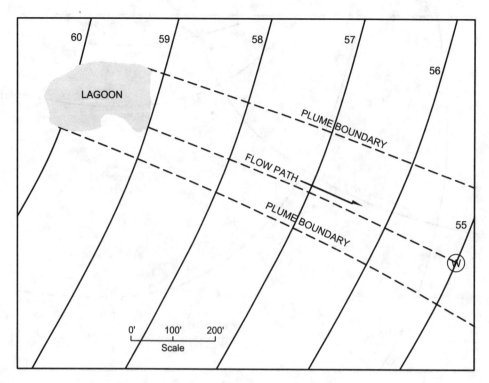

FIGURE 2-C
Flow path and plume extent for Exercise 3

As discussed in Note c at the beginning of these exercises, you may ignore diffusion of contaminated material out the sides of the plume, assuming instead that all the contamination stays in the portion of the aquifer that passes under at least some part of the lagoon, and that the concentration is uniform therein. Express your concentration results in ppm units (for which you may wish to review the material at the end of Section 2.4). The following sequence of individual questions may help you analyze this problem:

a)   How long will it take for the contamination to reach W? Analyze the flow path from the lagoon to W as you have done earlier.

The answer would be about 197 days, based on Figure 2-C and Table 2-E. (Since the hydraulic gradient is very uniform in this case, it would also be acceptable to treat the flow path as a single step.)

b)   What is the cross-sectional area of the contaminated portion of the aquifer? (This refers to the cross section relevant to Darcy's law calculations.)

TABLE 2-E
Spreadsheet for Exercise 3

GROUND WATER TRAVEL TIME FOR EXERCISE 3, SECTION 2.7									
Conductivity =		10 ft/day							
Porosity =		0.3							
Segment	Length	h1	h2	i(calc)	i(extrap)	i	v (ft/day)	t(days)	t(years)
1	202	59	58	0.005		0.005	3.768	54	0.1
2	297	58	57	0.005		0.005	3.865	51	0.1
3	293	57	56	0.005		0.005	3.946	49	0.1
4	182	56	55	0.006		0.006	4.187	43	0.1
Total								197	0.5

The plume widens very slightly, as shown by the dashed lines, which represent the outer borders of the part of the flow that "passes under at least some portion of the lagoon." Thus the cross-sectional area varies slightly as well. An "average" value would be a value about halfway downstream, where the plume is about 235 feet wide, based on the map scale. Since the thickness is given as 20 feet, the cross-sectional area would be $20 \times 235 = 4700$ ft^2.

c)   What is the volumetric flow rate in ft^3/day for this portion of the aquifer?

The hydraulic gradient values are easily seen to be very uniform at about .005. (The .006 calculated in the table is actually rounded up from about .0055.) Thus Darcy's law yields:

$$Q = KiA = 190 \times 0.005 \times 4700 = 4465 \text{ ft}^3/\text{day}.$$

d)   If roughly 50 cubic feet per day of cyanide solution are seeping into this aquifer, how much is this in units of pounds/day? (Assume normal water density for the solution as well.)

$$\frac{50 \text{ ft}^3}{\text{day}} \times \frac{62.4 \text{ lb}}{1 \text{ ft}^3} = 3120 \text{ lb/day}.$$

e)   Now, this is a 0.01% solution, so how many pounds/day of cyanide are entering the aquifer?

Since 0.01% is the decimal amount 0.0001, we have

$$3120 \times .0001 = .312 \text{ lb cyanide per day}.$$

f)   What is the concentration of cyanide, expressed in ppm, in the ground water at the time it leaves the lagoon site or reaches W?

Since we are ignoring further spreading of the plume along the flow pathway, the concentration will be the same when it arrives at the well as when it leaves the lagoon area. We have:

$$\frac{.312 \text{ lb cyanide/day}}{4465 \text{ ft}^3 \text{ water/day}} \times \frac{1 \text{ ft}^3 \text{ water}}{62.4 \text{ lb water}} \times \frac{1,000,000 \text{ lb water}}{1 \text{ million lb water}} = \frac{1.1 \text{ lb cyanide}}{1 \text{ million lb water}} = 1.1 \text{ ppm}.$$

Note that for low concentration solutions, as we have here both in the seepage into the aquifer and then in the aquifer itself, it is not necessary to distinguish between the amount of contaminant per amount of water or amount of contaminant per amount of solution, since the denominators in these two cases differ by a negligible amount.

4.   Suppose that a shallow aquifer averaging about 15 feet thick runs under the locale sketched in Figure 2-27. Its hydraulic conductivity is 50 ft/day, and its porosity is 30%. The contour lines show head values (in feet) in this aquifer. Also shown is an equipment repair facility ("plant") which, over the years, has been careless in its disposal of waste solvents and cleaning agents used to clean parts and equipment. In fact, the residue was generally dumped on the ground "to evaporate." Unfortunately, a substantial amount seeped into the ground before it could evaporate. Based on records of chemical purchases over the years, a consulting engineering firm determines that a long-term average of about two gallons per week were dumped out to evaporate and that about a quarter of this is estimated to have actually seeped into the ground. This has been taking place for at least the past 30 years. To test their understanding of the situation, they decide to sink test wells into the aquifer directly along the centerline of the plume along both the 65- and 75-foot contours.

a)   What concentrations should they expect to find in these two test wells (expressed in ppm)?

b)   What would be the expected travel time for contaminated water to move from the center of the plant to the test well on the 65-foot contour line?

Assume that the materials of interest here are water-soluble and that they just move along with the water at the same rate. This is a simplifying assumption that applies to some such agents but not to others. Assume further that the density of the waste liquids is essentially the same as that of water.

This problem is similar to the earlier one involving cyanide wastes from a gold mine, but the reader is not specifically told this, nor is the reader led through the solution step by step, as was done before. However, the

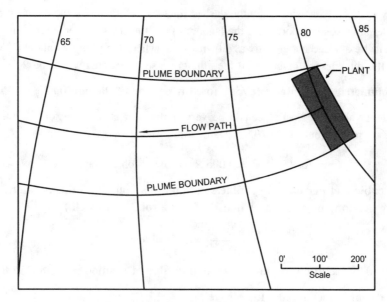

FIGURE 2-D
Flow path and plume extent for Exercise 4

solution should proceed along the same lines. The plume and flow centerline are shown on Figure 2-D, and the travel-time calculations are summarized in Table 2-F.

Thus the answer to part b would be about 292 days.

For the concentration calculations, the flow in the aquifer (using a cross section about halfway downstream, where the width is estimated to be 330′) is given by:

$$Q = KiA = 50 \times 0.0175 \times (330 \times 15) = 4,331 \text{ ft}^3/\text{day}.$$

Now we work out the concentration using our convenient unit conversion scheme:

$$\frac{.5 \text{ gal(solvent)/week}}{4,331 \text{ ft}^3\text{(water)/day}} \times \frac{1 \text{ week}}{7 \text{ days}} \times \frac{231 \text{ in}^3\text{(solvent)}}{1 \text{ gal(solvent)}} \times \frac{1 \text{ ft}^3}{12^3 \text{ in}^3} \times \frac{62.4 \text{ lbs(solvent)}}{1 \text{ ft}^3\text{(solvent)}}$$

$$\times \frac{1 \text{ ft}^3\text{(water)}}{62.4 \text{ lbs(water)}} \times \frac{10^6 \text{ lbs(water)}}{1 \text{ million lbs(water)}}$$

$$= \frac{2.2 \text{ lbs(solvent)}}{1 \text{ million lbs(water)}} = 2.2 \text{ ppm}$$

The conversion factor of 231 in the above was found in the dictionary under "gallon."

TABLE 2-F
Spreadsheet for Exercise 4

GROUND WATER TRAVEL TIME FOR EXERCISE 4, SECTION 2.7							
Conductivity =	50 ft/day						
Porosity =	0.3						
Segment	Length	h1	h2	i	v (ft/day)	t(days)	t(years)
1	269	80	75	0.019	3.10	86.8	0.24
2	292	75	70	0.017	2.85	102.6	0.28
3	292	70	65	0.017	2.85	102.6	0.28
Total						292.0	0.80

This next problem is a good problem for introduction in class, allowing the students to work in groups to try to solve it. After a while, if they don't get it, you can give them a hint something like: "Think about traveling by canoe or small boat downstream on a river. What kinds of things might happen as the river passes through wide and narrow parts, shallow and deep parts?" In presenting discussion problems like this, it can be useful not only to obtain a correct answer, but also to emphasize the importance of explaining it in a clear and convincing way. Thus sometimes after the answer has more or less been found, each group might have a chance to present the most convincing and clear explanation they can think of. The teacher can even prescribe a time limit (e.g., two minutes) or number of sentences, in order to get them to focus on the quality of what they say. Students have found this quite challenging and engaging in class.

5. **Consider the situation described in Figure 2-28, which shows a lined landfill with a leaky liner that is underlain by a shallow sand aquifer with head contours (in feet) as indicated. Suppose that the aquifer is uniformly 30 feet thick, its hydraulic conductivity is 10 ft/day, and its porosity is 0.4. Why is this situation, as described here, self-contradictory and hence impossible? (Hint: begin by carefully sketching the plume of ground water contaminated by the landfill.)**

In this situation, the contour lines tend to significantly "focus" the flow lines or the plume into a narrower one as you move down gradient, as can be seen in Figure 2-E. Since the aquifer is definitively stated to be uniformly thirty feet thick, conservation of mass would require that the water start to move faster as it goes downstream. However, this is inconsistent with the contour lines, which show an essentially uniform gradient. (The properties of the aquifer are uniform and hence cannot compensate.) To relate this to the river analogy suggested before the problem, think about canoeing down a river and coming to a narrow gorge. If the river stays the same depth, it will have to flow faster through the gorge to accommodate all the water coming into it. Equivalently, the only way the water could keep moving at the same speed would be if the water in the gorge got quite deep so as to fully compensate for the loss in width.

Note that it is certainly possible to have some focusing or convergence of flow lines, as we have seen in several examples, but this cannot occur when the hydrologic parameters are absolutely uniform and at the same time the

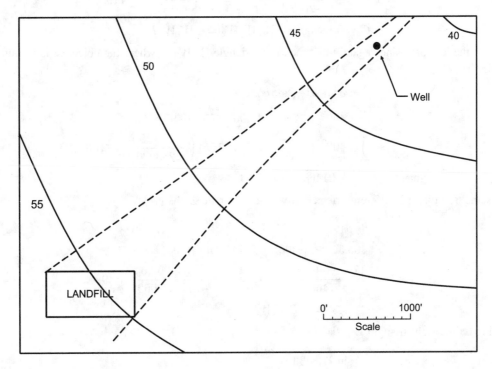

FIGURE 2-E
The apparent "focusing" or convergence of flow lines in Exercise 5

hydraulic gradient is also uniform. If the hydraulic head contour lines in this example were assigned values such that the gradient would be getting steeper in the region of convergence, then there would be no *a priori* basis for rejecting the plausibility of the situation. To anticipate the more advanced developments in Chapter 5, the set of all possible head distributions (for uniform aquifer parameters) is the set of solutions to Laplace's famous partial differential equation with boundary conditions appropriate to the ground-water problems under discussion.

Because local or regional variations in aquifer parameters are a dominant factor in causing variations and curvature in the head contours, the other examples and problems in this chapter generally use qualifiers such as "average" or "approximately" so as to avoid unwittingly constructing an inherently contradictory situation. But no such terminology was used in describing the situation in this exercise.

2.8 Determining Approximate Flow Directions Using Data from Three Wells

If you don't want to get involved in trigonometry, you can still do the second exercise, which is exactly analogous to the material in the text for this section. Or the trig problem could be used as an extra credit problem.

1. Use trigonometry to determine the exact length of the arrow discussed under step 6, above. This is the distance from the 20.9 foot contour line to Well 3. Use this new value to recalculate the hydraulic gradient.

The basic tool needed is the law of cosines, which we will always write in the same notation as "$c^2 = \ldots$," even as we apply it to varying triangles and orientations. All angle measures are in degrees, and much more precision is given than would be applied in a real situation just to make it easy to verify the calculations. Letting W_n stand for Well "n", we first solve the big triangle for $\angle W_1 W_3 W_2$:

$$c^2 = a^2 + b^2 - 2ab \cos C$$

$$575^2 = 400^2 + 350^2 - 2(400)(350) \cos C$$

$$\cos C = \frac{575^2 - 400^2 - 350^2}{-2(400)(350)} = \frac{48125}{-280000} = -.1719$$

$$C = 99.8969 = \angle W_1 W_3 W_2 = \angle W_1 W_3 P.$$

Next we apply the law of cosines again, this time to the triangle $W_1 W_3 P$, where the objective is to find the length of the midline $W_1 P$:

$$c^2 = a^2 + b^2 - 2ab \cos C$$

$$c^2 = 400^2 + 175^2 - 2(400)(175) \cos 99.8969$$

$$c^2 = 400^2 + 175^2 - 2(400)(175) \left(-\frac{48125}{280000} \right) = 214687.5$$

$$\text{segment } W_1 P = \sqrt{214687.5} = 463.3438.$$

Using this same triangle in a different orientation we can compute $\angle P W_1 W_3$:

$$c^2 = a^2 + b^2 - 2ab \cos C$$

$$175^2 = 463.3438^2 + 400^2 - 2(463.3438)(400) \cos C$$

$$\cos C = \frac{175^2 - 463.3438^2 - 400^2}{-2(463.3438)(400)} = \frac{-344062.5}{-370675.0598} = .9282$$

$$C = 21.8433 = \angle P W_1 W_3.$$

Now focusing on the right triangle $W_1 Q W_3$, we have:

$$\sin 21.8433 = \frac{QW_3}{400}$$

$$QW_3 = 400 \times .3721 = 148.8275$$

The recalculated gradient value would then be:

$$i = \frac{37 - 34}{148.8275} = .0202$$

which is essentially the same value as was obtained earlier by the method in the text.

2. Consider the situation shown in Figure 2-31, which is analogous to the example treated above. Determine the general direction of flow and the corresponding hydraulic gradient.

See Figure 2-F, the only real difference from the example in the text being that the point Q is outside the original triangle. As earlier, we find the length of QW_2 by use of the map scale or by repeated applications of the law of cosines (worked out here). To begin, we find $\angle W_1 W_2 W_3$:

$$c^2 = a^2 + b^2 - 2ab \cos C$$

$$180^2 = 210^2 + 260^2 - 2(210)(260) \cos C$$

$$\cos C = \frac{180^2 - 210^2 - 260^2}{-2(210)(260)} = \frac{-79,300}{-109,200} = .7262$$

$$C = 43.432 = \angle W_1 W_2 W_3 = \angle P W_2 W_3.$$

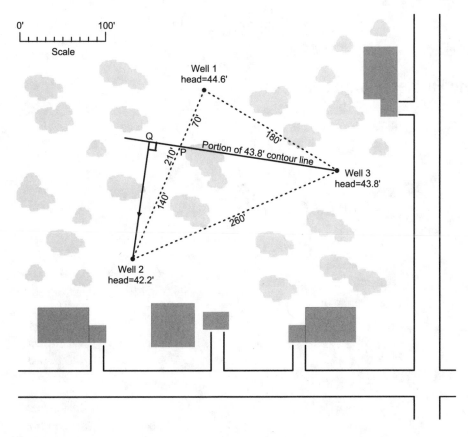

FIGURE 2-F
Framework for the solution of Exercise 2

Next we apply the law of cosines again, this time to the triangle PW_2W_3, where the objective is to find the length of the midline W_3P:

$$c^2 = a^2 + b^2 - 2ab \cos C$$

$$c^2 = 140^2 + 260^2 - 2(140)(260) \cos 43.432$$

$$c^2 = 140^2 + 260^2 - 2(140)(260) \left(\frac{79,300}{109,200} \right) = 34,333$$

segment $W_3P = \sqrt{34,333} = 185.293.$

Using this same triangle in a different orientation we can compute $\angle W_2PW_3$:

$$c^2 = a^2 + b^2 - 2ab \cos C$$

$$260^2 = 140^2 + 185.293^2 - 2(140)(185.293) \cos C$$

$$\cos C = \frac{260^2 - 140^2 - 185.293^2}{-2(140)(185.293)} = \frac{13,667}{-51,881} = -.2634$$

$$C = 105.273 = \angle W_2PW_3.$$

Now, focusing on the right triangle PQW_2, one base angle is the supplement of the angle just calculated, so that:

$$\angle QPW_2 = 180 - 105.273 = 74.727$$

$$\sin 74.727 = \frac{QW_2}{140}$$

$$QW_2 = 140 \times .9647 = 135.0554.$$

The recalculated gradient value would then be:

$$i = \frac{43.8 - 42.2}{135.0554} = .0118.$$

2.9 Guide to Further Information

(No exercises)

3

Air Quality Modeling

3.1 Background

[Note: several of these exercises ask you to target your response to a specific type of imaginary audience. You are being asked to play a specific role, and you should be sure to put yourself in this role as you prepare your response. Remember that just collecting the appropriate information is not enough; you must be able to communicate it effectively and efficiently to whatever audience you are dealing with.]

1. You recently got a job at Ace Consulting. They have expanded their services to include air quality analysis. The last set of major changes to the Clean Air Act has their clients confused and calling daily for some help in understanding exactly what they need to do to comply. Your boss tells you that you are to participate in a workshop for these clients. Your assignment will be to introduce the program by giving a very brief synopsis of the Clean Air Act. You'll need to include its basic structure (e.g., the classes of pollutants being addressed) as well as how this law has been updated since its original enactment. You're told you need to submit a two-page summary of your introduction to be included with the papers given by the other associates. Here's your chance to shine. You need to show good judgment in picking information that will give both the "big picture" as well as key specifics. Target your audience and write the summary.

Note: This response is not specifically written in the role described in the problem; it is intended to give the teacher the background information necessary to judge such responses.

Major Federal legislation generally requires periodic review or "reauthorization," with changes, if any, taking the form of an amended act. The Clean Air Act dates back to 1963, but the present regulatory structure dates to the 1970 and 1977 amendments. This general structure includes a focus both on ambient air quality (National Ambient Air Quality Standards) and on regulations that more directly affect individual sources (New Source Performance Standards, and the Prevention of Significant Deterioration principles). The principal pollutants addressed prior to the 1990 amendments were the so-called "criteria pollutants" (particulates, sulfur dioxide, carbon monoxide, nitrogen oxides, ozone, and, more recently, lead), for which ambient standards were issued, as well as eight toxic substances (arsenic, asbestos, benzene, beryllium, coke oven emissions, mercury, radionuclides, and vinyl chloride) for which emissions standards were issued. Shortcomings of this system included poor levels of achievement of ambient standards, especially for ozone, regulatory loopholes (such as waivers for tall stacks) that aggravated problems such as acid rain, and the failure of regulators to develop emissions standards for other important toxic substances.

The 1990 amendments (the most recent overhaul at the time of this writing) tried to overcome these difficulties by putting greater pressure both on polluters and on the states, who in large measure are responsible for implementation of the requirements. Examples of performance and emissions requirements give a flavor of the some of the regulatory concepts: NSPS (new source performance standards), NESHAPS (national emission standards for hazardous air pollutants), BACT (best available control technology), BART (best available retrofit technology), RACT (reasonably available control technology), LAER (lowest achievable emission rate), and FMVCP (federal motor vehicle control program). (Pressure on states who are somewhat recalcitrant in implementing pollution control measures of the act can be brought in a variety of ways, such as by the withholding of Federal highway funds.) In addition to this greater emphasis on helping the states implement more effectively the ongoing areas of concern under the act, key changes implemented in 1990 include:

- Identifying 189 toxic air pollutants for inclusion on the regulatory agenda for emissions standards;

- Increasingly tightened vehicle emissions standards, as well as regulatory authority to control emissions of other equipment, such as motorboats, lawn mowers, etc.;

- Gasoline modifications for areas with carbon monoxide problems, to promote more complete combustion;

- Required vehicle emission inspection programs in areas with air problems;

- Overall reductions in sulfur emissions from coal-burning power plants (to combat acid rain), as well as nitrogen oxide emissions;

- Regulatory attention directed to estimating health risks to people from air toxins, including materials released as a result of accidents;

- Allowances for economic incentive schemes for maximizing the impact of pollution control investments (e.g., trading "emission allowances");

- Stratospheric ozone protection by control and phase-out of CFCs and related substances.

As an appendix to this answer, especially to enable the teacher to see how different student responses fit in with the overall framework, the Table of Contents of the actual Act, below, may give a useful overview. (Often it is more beneficial and not that daunting to look at original sources rather than summaries.) Both full and "plain English" summaries of the act are maintained on the EPA's on-line service. Regulations are found in the Code of Federal Regulations, by agency; and the EPA's regulations are in Volume 40, available in most libraries and from many on-line sources. Another excellent source for material such as this would be major newspaper or news magazine stories associated with new amendments.

TITLE I - AIR POLLUTION PREVENTION AND CONTROL

Part A - Air Quality and Emission Limitations

Sec. 101. Findings and purposes.
Sec. 102. Cooperative activities and uniform laws.
Sec. 103. Research, investigation, training, and other activities.
Sec. 104. Research relating to fuels and vehicles.
Sec. 105. Grants for support of air pollution planning and control programs.
Sec. 106. Interstate air quality agencies or commissions.
Sec. 107. Air quality control regions.
Sec. 108. Air quality criteria and control techniques.
Sec. 109. National ambient air quality standards.
Sec. 110. Implementation plans.
Sec. 111. Standards of performance for new stationary sources.
Sec. 112. National emission standards for hazardous air pollutants.
Sec. 113. Federal Enforcement.

Sec. 114. Inspections, monitoring, and entry.
Sec. 115. International air pollution.
Sec. 116. Retention of state authority.
Sec. 117. President's air quality advisory board and advisory committees.
Sec. 118. Control of pollution from federal facilities.
Sec. 119. Primary nonferrous smelter orders.
Sec. 120. Noncompliance penalty.
Sec. 121. Consultation.
Sec. 122. Listing of certain unregulated pollutants.
Sec. 123. Stack heights.
Sec. 124. Assurance of adequacy of state plans.
Sec. 125. Measures to prevent economic disruption or unemployment.
Sec. 126. Interstate pollution abatement.
Sec. 127. Public notification.
Sec. 128. State boards.

Part B - Ozone Protection

The 1990 CAA Amendments replaced Part B with Title VI - Stratospheric Ozone Protection, which follows later.
The following sections of Part B were replaced:

Sec. 150. Purposes.
Sec. 151. Findings and definitions.
Sec. 152. Definitions.
Sec. 153. Studies by environmental protection agency.
Sec. 154. Research and monitoring by other agencies.

Sec. 155. Progress of regulation.
Sec. 156. International cooperation.
Sec. 157. Regulations.
Sec. 158. Other provisions unaffected.
Sec. 159. State authority.

Sec. 402. Definitions.

Sec. 403. Sulfur dioxide allowance program for existing and new units.

Sec. 404. Phase I sulfur dioxide requirements.

Sec. 405. Phase II sulfur dioxide requirements.

Sec. 406. Allowances for States with emissions rates at or below 0.80lbs/mmBtu.

Sec. 407. Nitrogen oxides emission reduction program.

Sec. 408. Permits and compliance plans.

Sec. 410. Election for additional sources.

Sec. 411. Excess emissions penalty.

Sec. 412. Monitoring, reporting, and recordkeeping requirements.

Sec. 413. General compliance with other provisions.

Sec. 414. Enforcement.

Sec. 415. Clean coal technology regulatory incentives.

Sec. 416. Contingency guarantee; auctions, reserve.

TITLE V - PERMITS

Sec. 501. Definitions.

Sec. 502. Permit programs.

Sec. 503. Permit applications.

Sec. 504. Permit requirements and conditions.

Sec. 505. Notification to Administrator and contiguous States.

Sec. 506. Other authorities.

Sec. 507. Small business stationary source technical and environmental compliance assistance program.

TITLE VI - STRATOSPHERIC OZONE PROTECTION

Sec. 601. Definitions.

Sec. 602. Listing of class I and class II substances.

Sec. 603. Monitoring and reporting requirements.

Sec. 604. Phase-out of production and consump- tion of class I substances.

Sec. 605. Phase-out of production and consump- tion of class II substances.

Sec. 606. Accelerated schedule.

Sec. 607. Exchanges. [Exchange authority.]

Sec. 608. National recycling and emission reduction program.

Sec. 609. Servicing of motor vehicle air conditioners.

Sec. 610. Nonessential products containing chlorofluorocarbons.

Sec. 611. Labeling.

Sec. 612. Safe alternatives policy.

Sec. 613. Federal procurement.

Sec. 614. Relationship to other law.

Sec. 615. Authority of Administrator.

Sec. 616. Transfers among Parties to the Montreal Protocol.

Sec. 617. International cooperation.

Sec. 618. Miscellaneous [provisions].

2.   After college you wanted some international experience, and so you took a position in the international division of Hirosaki Industries in Japan. The company's business is building and operating trash incinerators that also usually generate electricity. The Board of Directors has suggested expanding from current markets in Asia and the Pacific into the US, and your Vice-President sends you to the US for a couple of weeks to find out what problems have occurred in permitting or operating trash incinerators there in the last ten years. You wind up the field part of your assignment and are about to take a leisurely trip back with a stopover for some much needed R&R in Hawaii when you get an urgent call from his secretary asking you to be back to meet with him in his office at 9 A.M. the next day, bringing along a memo summarizing your observations. He has apparently been asked to give an interim briefing to the board that evening. Nothing new, just another "crunch" moment you can expect in any new job! (You're the low person on the ladder, and you're used to them.) You race to airport for an all night flight, and along the way you type your memo on your notebook computer. You know that the boss likes facts, he likes graphic organization, and he likes brevity. One page maximum. Write the memo. [Note: assume that the US visit focuses either on the region where you live or on the region where your school is located, the term "region" referring to several states. Make good use of your library's computerized databases and other sources. Computerized newspaper files can also be an excellent source of information, especially when there is controversy over some issue.]

As with the previous exercise, this response is not being written as a sample answer, but rather as a source of information for the teacher. There are about 130 municipal solid waste (MSW) incinerators in the US, handling about one-sixth of the MSW waste stream. Of these, most also generate electricity, and any new one would almost certainly include electricity generation as part of its design. In addition, most recent plants also include some recycling, such as metals recovery before or after burning, or the use of a portion of the ash as aggregate for construction.

The most significant recent problem with operating facilities in many locations has been trash availability, which may be surprising. In particular, these facilities are usually built by private companies under long-term contracts with municipalities: these municipalities guarantee a certain amount of waste, and, in turn, are guaranteed a certain cost structure for the disposal of the wastes. This gives reasonable stability to both parties. However, fairly extensive recycling efforts have cut heavily into trash volume in some states and regions, so that municipalities are no longer able to meet their quotas of trash. This can raise the per ton disposal cost considerably, and it is generally not offset by recycling income, which can be quite meager if even positive. Thus some communities are eagerly trying to import trash, which sounds terrible to other communities who may be thinking about a future incinerator.

Other problems for which there is abundant documentation in the press include odors, landfilling of ash that contains heavy metals, toxic air emissions, and heavy truck traffic. With respect to odors, most facilities operate under negative pressure, meaning that when the doors are opened for a truck to bring in a load of trash, the air from the outside gets sucked in, rather than vice versa. This negative pressure is created by drawing the air from inside into the combustion chamber to support the combustion, where the odoriferous components would also be destroyed. But this is not always effective, and, in addition, when the facility is down for maintenance or repairs, the residual trash can cause a real neighborhood nuisance.

Concerns over toxic ash are based primarily on the leachability of heavy metals, and whether this is of real concern depends on the design of the landfill where the ash is being deposited. Many modern landfills have both liners and leachate collection systems at the bottom, as well as reasonably impervious caps on top intended to prevent water from percolating through. Air pollution issues relate to both criteria pollutants and toxics, the latter including certain metals (such as mercury from batteries), possibly dioxin (from the burning of plastics), and other more or less unpredictable materials due to the somewhat uncontrolled nature of the fuel source—trash. Pollution control equipment is effective in removing a large portion of the substances of concern, but even if it is 99% effective , say, that means that 1% is getting through. This invariably causes concerns to residents living near such a facility.

The case of trash incinerators is an interesting one to discuss in an environmental class because it demonstrates how the partial solution of one environmental problem (space for and control of landfilling) can lead to others (e.g., air pollution). It also helps to add a realistic perspective on issues like recycling, where success depends not only on long-term philosophical merit, but also on being able to overcome practical and political issues such as market structure and trash disposal costs.

While it does not fit the air pollution theme of this chapter too directly, as a follow-on to the previous exercise it can be interesting to ask members of the class if they know where the trash from their homes goes. (Most of them usually don't know, especially if they are from cities.) This can lead to a further assignment or extra credit problem in the form of tracking down and diagramming the waste stream(s), including any separations or recycling. Alternatively, they could be asked to track the college's or other institution's various waste streams, of which there can be many.

3. Your state environmental agency probably has separate divisions responsible for different aspects of the environment, such as air, ground water, surface water, etc. As part of an ongoing election campaign, a prominent candidate promises that she will be able to reduce taxes by cutting back on unnecessary government operations. One of the agencies that she has proposed to cut back is the state environmental agency, with particular emphasis on the large contingent of staff responsible for air pollution work. Her position is that air pollution problems have been brought largely under control and that there is no longer nearly the same need for a large staff that there might have been a decade or two ago. You are working as a summer intern with a nonprofit citizens organization that is deciding which candidates to endorse for election, and you have been asked to prepare a briefing paper for the board of directors concerning the validity of this candidate's claims about the level of staffing. By contacting the agency and/or using publicly available reference material, find and present clearly and succinctly the following information: a general organization chart of the portion of the environmental agency that deals with air pollution, a breakdown of the staffing of this branch and its overall budget level, and the three most important activities of this part of the agency within the last five years.

This answer is, of course, state-specific. Some states (e.g., Massachusetts) provide essentially all of the required information on-line, although the budget information may need to be obtained from a different part of the state's database than the part related to the activities and organization of the environmental agency. This is all public information, and electronic or telephone inquiries will generally lead to the desired information as well, as will consultation of library resources. Regarding the last part of the question, annual reports or published speeches of agency officials usually emphasize what they consider to be the major agency accomplishments or activities, since this kind of "promotion" is part of the political process. (But remember that since they are indeed promoting their activity and justifying their existence, one should apply a critical perspective in reviewing such material.)

4.   This problem applies to the particular school or other institution with which you are associated. Find out who within the organization is responsible for environmental permits, and determine whether there are any such permits related to air pollution. If there are, compile a general list of emission sources for which such permits exist, and examine one individual permit, preparing a detailed summary of its content.

A typical permit for a boiler might regulate the opacity of the exhaust and specify fuel type and operating conditions, which are related to other kinds of emissions. This might be found in the physical plant department or in some other administrative office. Laboratory hood exhausts and other items might also be controlled by air permits. Large or complex plants might regulate additional parameters.

5.   The text of this section referred to the occurrence of a number of incidents in the United States during which deaths and illnesses were caused by air pollution. Find an example of one such incident and provide a brief summary of what happened.

Two incidents referred to in the *Environmental Almanac* (compiled by the World Resources Institute) are: 80 deaths in New York City from an air pollution incident in 1966 in which an atmospheric inversion (see below) for four days led to the buildup of pollution; 20 deaths and an estimated 14,000 cases of illness in Donora, Pennsylvania in 1948, again due to an atmospheric inversion. A search of the *New York Times* for 1966 provided these details on the New York City incident: the problems occurred in a three-state area (New York City and parts of New Jersey and Connecticut) over the Thanksgiving weekend and made first-page headlines every day. Politicians' commentaries were somewhat caught up in politics, as a vigorous debate was going on about proposed new and expensive city regulations on air pollution sources. Initial comments by the health commissioner argued that there were no additional deaths or serious illnesses caused by the incident, an assertion that was no doubt later proved false by statistical analysis of the data. One article also referred back to a Thanksgiving Day 1953 smog emergency in the city, to which 240 deaths had been attributed. An on-line search under "air pollution disaster" provided the following additional information about the Donora incident. Twenty people died and approximately 7,000 or 50% of the population, experienced acute illness during the week of Oct. 25, 1948, when temperature inversion and air stagnation occurred. Persons of all ages became ill, but those over 55 were more severely affected. Those with previous heart or respiratory disease, particularly bronchial asthma, suffered most. Symptoms were primarily respiratory and secondarily gastrointestinal, and included cough, sore throat, chest constriction, shortness of breath, eye irritation, nausea, and vomiting. The onset of the illness for most persons occurred on the evening of the third day. Of the 20 who died, 14 had some known heart or lung disease.

An atmospheric inversion may be described as follows. Under normal conditions, the air is warmest at ground level and is cooler as you move higher and higher. Since warm air rises due to buoyancy, this tends to carry away or otherwise dilute pollutants at or near ground level. An inversion is a meteorological condition in which the air is warmer at greater heights, at least in some layer, and so there is no tendency for the air to rise from ground level. The so-called inversion layer serves as a cap on any polluted air below, and thus the pollution can become quite stifling. From the above newspaper accounts, measurements taken at 7 A.M. in New York City on Friday, November 25, showed a surface temperature of 42°F and a temperature at 1,500′ altitude of 53.5°F, precisely this kind of inversion situation.

Outside the US, but still in the developed world, London has experienced repeated incidents of this type, with one in 1952 estimated to have caused 4,000 deaths. Aside from localized incidents, it is also important to note that air pollution can cause so-called "statistical deaths," that is, a general increase in the number of deaths over and above what would be expected, even if the specific individuals involved cannot be identified. For example, the American Lung Association reported in 1996 its estimate that about 2,000 premature deaths could be prevented in the US if the entire nation were simply to adopt then current California standards that control particulate emissions.

Particulate pollution is especially harmful to people with asthma, emphysema, and chronic bronchitis. They further estimate that in total over 120,000 premature deaths are caused by air pollution each year in the US. (Information is readily available on-line.) To a certain extent, even the death estimates given above for the particular incidents are based on statistical analysis of perturbations from the expected rate, but they are based on identifiable incidents and target populations. The statistical deaths referred to in this paragraph are somewhat different in that they are usually based on dose-response curves obtained by a variety of methods, combined with calculations of exposure levels on a national basis.

6.   For both your own hometown and your institution's location, determine whether they are located in a "non-attainment area" for ozone, meaning that the national air quality standard for this pollutant is not being met. If so, determine the degree of severity of non-attainment according to the EPA's classification scheme. (Hint: this can be determined on-line by accessing and searching EPA's resources, or it can be determined by review of published reports or by making inquiries to environmental organizations or agencies.)

Following the hint, one is easily led both to maps and searchable lists of nonattainment areas for all of the criteria pollutants, including ozone. (This refers to tropospheric ozone associated with smog, not stratospheric ozone related to protection from ultraviolet rays of the sun.) The number of people living in such zones is considerable, as indicated in the following chart from the on-line EPA database:

Ozone Nonattainment Area Summary (As of December 9, 1996)

Classification	Number of areas	Number of counties	Population (1000s)
Extreme	1	4	13,000
Severe	9	78	42,056
Serious	11	76	22,961
Moderate	19	63	18,231
Marginal	26	62	10,599
SUBTOTAL	66	283	106,847
Transitional	5	11	2,532
Incomplete data	21	24	1,842
GRAND TOTAL	92	318	111,221

3.2 Physical Principles

1.   Give an example of a diffusion process with which you are familiar but other than those discussed in the text. Identify whether this situation involves one-, two-, or three-dimensional diffusion.

Bomb- or drug-sniffing dogs detect low concentrations of molecules in the air that have diffused away from their sources. This would generally be three-dimensional and may occur in conjunction with other processes, such as advection (being carried along by moving air). A puff of cigarette smoke in the air gradually disperses, even if the air is very still, at least in part by three-dimensional diffusion. (There is also some gravity settling of heavier particles in the smoke.) If you open a can of paint in a relatively still room, the solvent vapors spread by three-dimensional diffusion. If you paint a wall, the evaporating solvent vapors spread by one-dimensional diffusion in the direction perpendicular to the wall (although of course the process would be "one-sided" in that material could only move into the room and not out the other side of the wall). Intermediate between these cases, if you were to paint a post or column in the center of the room, say, the vapors would disperse by two-dimensional diffusion in the two horizontal directions. A spill of soluble material into the center of a large shallow lake would lead to dispersal by two-dimensional diffusion. Precisely planned diffusion processes also play a key role in mixing chemical layers in the development of instant photography film.

2.   In the three-dimensional example of the diffusion of evaporating perfume through a room, imagine that you modify the situation by turning on a small fan in one corner of the room. The fan is not aimed at the tissue containing the perfume, but rather it is aimed along the wall on one side of the room. Would this increase or decrease the rate at which perfume disperses throughout the room? Explain your answer.

It would increase the rate of dispersion because it would add an additional mixing process, advection, to the system, as well as increasing the diffusion rate itself because of the turbulence it would create in the air medium.

3.   Suppose that exhaust gases are leaving a vertical smoke stack at a constant rate while the wind at the top of the stack is blowing at 10 miles per hour. There will be a certain effective height of release H somewhat larger than the actual physical height of the stack h, as discussed in the text. Now suppose that the wind speed is increased to 20 miles per hour. What effect do you expect this to have on the effective stack height H? Explain your answer.

It should decrease the effective stack height because now the same amount of exhaust is being mixed with twice as much air upon release from the stack, so the dilution effect on both heat and momentum of the exhaust gases should be greater and faster. The precise amount cannot be determined from the problem. See Exercise 9 in Section 3.7 for an introduction to the quantitative aspects of this issue.

4.   As you know, gases from an exhaust plume can certainly disperse both horizontally and vertically. For gases that are dispersing in a downward direction, what do you suppose happens to them when they reach the actual ground level? For example—keeping in mind that for diffusion processes such as this, the net movement of materials is always from an area of higher concentration to lower concentration—could there ever be a layer just above the ground where the pollution would build up to a higher concentration than that just above it? Be very precise in your explanation.

> This question can be treated on either a simple intuitive level or on a more advanced one, depending on the background of the students. Even well-trained math students can start making fallacious inferences when trying to give a precise logical argument to back up their opinion, and this can be a good teaching opportunity. Here we emphasize the simple analysis, which itself has a twist that many overlook, relegating a more advanced investigation to Chapter 6.

The gaseous pollutants start to accumulate near the ground, but not at a level that would cause a rise in concentration there higher than the overlying layers. This is due to the fact that as the material starts to accumulate just above the ground, this causes the concentration gradient to get smaller, and the process is "self-limiting" because any time the gradient might possibly get all the way to zero, no more material could get by that point (to start to build up an even higher value on the other side).

However, this behavior applies only **until** well into the diffusion process when essentially everything eventually has to diffuse off in the positive vertical direction. Since every single molecule eventually moves up and away, even if it initially moved down towards the ground, for every point at every single height above the ground, at some time the concentration gradient there will have to be negative in the upward direction (or else material would not move that way), meaning higher concentrations closer to the ground and hence a reversal of the concentration profile observed in the earlier moments of the process.

For beginning students or those with limited mathematical background, I think that the emphasis is properly on the first part of the explanation above, even if they fail to note the "**until**" part. Much depends on the level of the class and the objectives of the course.

Comment for classes with advanced students: Good students (and even faculty) often miss the "**until**" part above, and then the teacher, without hinting that anything is wrong, can press them to develop a solid logical argument to prove this. Some who are trained in proofs, might even use real analysis arguments (e.g., "look at the infimum of the set of times for which there is a point such that. . . "). But these arguments must invariably fail, and it is a good experience for these kinds of students to see how they fail and to see how they themselves may have misconceptions about how local maxima can evolve in smoothly changing families of curves. Once one reaches this point, it is even possible to go further and ask them to formulate a more limited theorem that captures the original situation they had envisioned in which you don't expect a local minimum to develop between the source point and the ground, at least while the local maximum associated with the source point has not moved too far.

5. Suppose you were trying to calculate the concentration of a pollutant at your residence, which is located exactly one mile downwind and in a direct line from a smoke stack that is 200 feet high. Here you know the value of h but you do not know how to calculate the effective stack height H. Suppose you used the value of h instead of H for your analysis. Would you be overestimating or underestimating the amount of pollution at your residence? Explain your answer carefully.

You would be overestimating the concentration at the residence because the axis of the plume in this case would be closer to the ground and all other conditions would be the same.

6. [Library research question.] Find and explain the precise scientific definition of atmospheric stability. (Hint: Look in a meteorology or air pollution text, or in some more general science or reference source, or consult on-line computer sources.)

This stability is similar to the concept of the stability of an equilibrium point, with which you (although not your students) are likely familiar. In particular, the point is stable if the system wants to return to the point if slightly perturbed. The point is unstable if a slight perturbation sends the system off in some new direction. The point is neutral if a slight perturbation stays, more or less, a slight perturbation. With respect to an air mass, consider a small portion of it at an arbitrary altitude, and think of it as being bounded by an imaginary balloon. Conditions are called stable if, when you push the balloon up a little, it wants to go back down, and if, when you push it down, it wants to go back up. Conditions are unstable if, when you push the balloon up a little, it wants to take off and go up higher; similarly when you push it down from the starting point. Conditions are neutral or neutrally stable if, when you push the balloon up or down, it just stays at the new location.

To develop this theme in greater detail for the atmospheric case, we need some simple preliminary facts: When you compress a gas, it heats up; for example, that's what causes ignition in a diesel engine. Conversely, when you expand a gas, it cools down. Both of these assume that heat transfer to the surroundings are negligible, in which case the process is called *adiabatic*. Furthermore, when the pressure on a gas is lowered, it expands; and this occurs for example, if a quantity of gas in the atmosphere rises to a higher altitude, for then it is under lower pressure (since it is under the weight of less atmosphere pushing down on it). Putting these two facts together, when a mass of air rises (adiabatically) in the atmosphere, it should experience lower pressure, hence expand, hence cool down. Using thermodynamics, one could even calculate the rate, which turns out to be about $-1°C$ per 100 meters of altitude. This is shown by the dashed line in the first graph in Figure 3-A. Note that only the slope of this line is fixed; if we intended to apply it to a given mass of air, we would want to have it pass through the actual starting temperature condition.

Now let us consider a graph of the actual measured temperature profile in the atmosphere, which need not bear any relation to this theoretical graph since it is affected by many additional local and regional weather factors. The second graph in the figure shows such a profile—in fact, one that even has the opposite slope: namely, higher temperature at higher altitude, which we previously referred to as an inversion condition. Look at what would happen if we were to take an arbitrary portion of this air mass, indicated by the smaller circle in the third graph, and perturb it slightly upward from its initial position. It would cool according to the theoretical curve shown in the first graph, the result being that it would now be colder than its surroundings, hence more dense, and therefore it would want to sink back down again. This denotes "stable" behavior, in the sense described earlier, and in fact this is called a stable atmospheric condition. It is easy to see that perturbations in the downward direction would similarly lead to a return in the direction of the starting point.

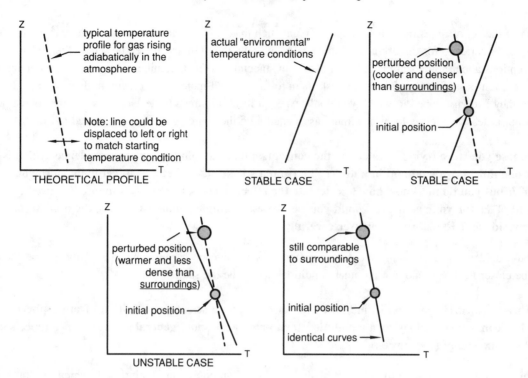

FIGURE 3-A
Illustration of principles of atmospheric stability and instability

To consider the opposite extreme, the fourth graph shows a measured temperature profile that falls off with altitude, even more rapidly than the dashed line. In this case, a slight upward perturbation of a small part of the mass does once again cause a drop in temperature, but not enough to keep up with the environmental drop in temperature at that new altitude. As a consequence, the small mass will now be warmer and hence less dense than its surroundings, and it will want to rise further because of its resulting buoyancy. This is typical "unstable" behavior, and indeed this is the definition of atmospheric instability.

On the borderline between the two cases is so-called "neutral stability," shown in the last graph, in which vertical perturbations of a small part of the air mass do not give it any further impetus to move up or down.

The atmospheric stability classes discussed in the text simply represent an attempt to relate readily observable field conditions to the expected form of the vertical temperature profile curve.

3.3 Typical Quantitative Issues

1. As has been mentioned earlier, when a large project is proposed that may have significant air emissions of one kind or another, it is generally required by the regulatory authorities that extensive air pollution calculations be carried out to predict the levels of concentrations of various pollutants in the vicinity of the facility. These environmental documents become matters of public record, and they are generally available for review by members of the public in the offices of the regulatory authority as well as quite often in town or city offices or local libraries in the vicinity of the proposed facility.

Identify one such project that has been analyzed in this way in your region or state. Find the location of available copies of the environmental studies, review these documents, and identify the key issues involving air pollution that needed to be resolved prior to making the decision on the acceptability of the project. Use secondary sources, such as newspaper accounts, if you do not have ready access to the primary sources. (However, you will learn much more about the context within which air pollution analysis is applied if you

can gain access to the original documents, which is usually not difficult.) Examples of the kinds of projects for which such analysis would generally be required include: power plants (both nuclear and fossil fuel plants), chemical manufacturing facilities, incinerators, some large landfills, radioactive or hazardous waste management facilities, cement plants, and major transportation projects. You may be able to use computerized or paper resources in your library to identify such projects, or consult on-line sources. Telephone calls to environmental agencies and local governmental authorities are also a good way to get started on a project such as this.

The answer to this question depends on where you are located, but it is not difficult to find this information. Depending on the nature of the plant or operation and the state in which it is located, the actual type of document may vary. For example, an Environmental Impact Statement (EIS) is a report required to be issued by a Federal agency prior to taking any action that may have significant environmental impacts. For example, major construction at a Federal facility or the issuance of a Federal approval for some major private action would generally involve the development of an EIS, usually by contractors working for the agency. Other types of documents may include Environmental Impact Reports, Environmental Assessments, Safety Analysis Reports, License Applications, and many others. Federal agency actions and reports are often discussed in on-line sources, thus a good place to try to identify a project and find the location of documents. The state agency responsible for air issues would also be a good source to turn to for suggested projects to review.

As examples, two very interesting cases in the Boston area include a proposed incinerator, located in the city, for Boston's trash, which was rejected (perhaps more for political reasons) after extensive air pollution modeling; and a co-generation power plant for the Longwood medical area, which was finally built after many years of study and controversy. After discussing reports on this latter case for homework and in class, this site has made for very interesting field trips, and air modeling issues continue to develop there. (For example, if some of the hospitals expand significantly, how will that affect air patterns and pollutant dispersion around the plant?)

2.   Identify the role of quantitative air pollution analysis in the development of at least one environmental regulation by the US Environmental Protection Agency (EPA). Note that when an environmental agency intends to promulgate new regulations, it goes through a number of rounds of draft regulations for public review and comment. Furthermore, draft and final regulations are usually preceded by a carefully constructed preamble that summarizes the kinds of studies that were carried out in support of the new regulations. It will not be difficult to find extensive information of the type sought here from many different sources, such as accessing the EPA on-line, reviewing copies of the Federal Register (the official federal document in which such new regulations or draft regulations are published), probably located in your library, or by using a variety of search tools, including newspaper indexes, available in your library.

There is much benefit to be gained by looking at some of this "real world" context of quantitative modeling before we get into the more detailed mathematical calculations in the sections that follow. The Federal Register is available on-line in searchable form, from which the following excerpt has been taken. It is only one of very many citations resulting from a search on the term "air modeling." This should give a flavor for the way the government develops regulations. It is only a partial excerpt of the full entry.

[Federal Register: December 13, 1996 (Volume 61, Number 241)] [Proposed Rules] [Page 65715-65750]
From the Federal Register Online via GPO Access

ENVIRONMENTAL PROTECTION AGENCY

40 CFR Part 50

National Ambient Air Quality Standards for Ozone: Proposed Decision

ACTION: Proposed rule.

SUMMARY: In accordance with sections 108 and 109 of the Clean Air Act (Act), EPA has reviewed the air quality criteria and national ambient air quality standards (NAAQS) for ozone and particulate matter (PM). Based on these reviews, the EPA proposes to change the standards for both classes of pollutants. This document describes EPA's proposed changes with respect to the NAAQS for ozone. The EPA's proposed actions with respect to PM are being proposed elsewhere in today's Federal Register.

The EPA conducted exposure analyses to estimate ozone exposures for the general population and two at-risk populations, "outdoor children" and "outdoor workers," living in nine representative US urban areas. The areas include a significant fraction of the US urban population, 41.7 million people, the largest areas with major ozone nonattainment problems, and areas that are in attainment with the current NAAQS. Exposure estimates were developed for a recent year, as well as for modeled air quality that simulated conditions associated with attainment of the current NAAQS and various alternative standards. The exposure analyses provide estimates of the size of at-risk populations exposed to various concentrations under different regulatory scenarios, as presented in section V.G of the Staff Paper and summarized below. These estimates are an important input to the risk assessment summarized in the next section. The probabilistic NAAQS exposure model for ozone (pNEM/OZONE) used in these analyses builds on earlier deterministic versions of NEM by modeling random processes within the exposure simulation. The pNEM/ OZONE model takes into account the most significant factors contributing to total human ozone exposure, including the temporal and spatial distribution of people and ozone concentrations throughout an urban area, the variation of ozone levels within each microenvironment, and the effects of exertion (which is represented by ventilation rate) on ozone uptake in exposed individuals. A more detailed description of pNEM/OZONE and its application is presented in section V.G of the Staff Paper and associated technical support documents (Johnson et al., 1994; Johnson et al., 1996 a,b; McCurdy, 1994a).

The regulatory scenarios examined in the exposure analyses include 1-hour ozone standards of 0.12 ppm (the current NAAQS) and 0.10 ppm, and 8-hour standards of 0.07, 0.08, and 0.09 ppm, the range of alternative 8-hour standards recommended in the Staff Paper and supported by CASAC as the appropriate range for consideration in this review. These analyses used 1- and 5-expected-exceedance forms of the standards and are based on use of a single year of data. These estimates were also used to roughly bound exposure estimates for other concentration-based forms of the standard under consideration (e.g., the second- and fifth-highest daily maximum 8-hour average ozone concentration, averaged over a 3-year period) by using air quality analyses that compare alternative forms of the standard, as presented in Section IV and Appendix A of the Staff Paper. The estimated exposures reflect what would be expected in a typical or average year in an area just attaining a given standard over a 3-year compliance period. Additional air quality and exposure analyses were done to estimate the exposures that would be expected in the worst year of a 3-year compliance period...

3. **Based on a review of newspaper articles or other information available in your library or on-line, identify what you would regard as:**

a) **the city in the United States that has the worst air pollution situation;**

b) **the city in the world that has the worst air pollution situation.**

Describe briefly each of these situations and explain the reason why you would characterize it as the worst in its class.

To give you a reasonably wide context to evaluate student responses, what follows in Table 3-A is a summary of 1992 data found on-line identifying the worst US cities with respect to the criteria pollutants. One can see that the South Coast Air Basin of California, the region around Los Angeles, is dominant in both ozone and nitrogen oxides, the dominant components of smog, and also has the worst city for two other pollutants. This is why Los Angeles usually carries the reputation as the major city with the worst air pollution in the US and also why California is generally regarded as leading the way for the development of new air pollution regulations, especially for vehicles, the source of many of these problems. Another on-line source states that the five US cities with the highest number of premature deaths from suspended particulates in a recent year were Los Angeles, where 5,873 annual deaths were attributed (by statistical analysis) to fine particulate pollution, followed by New York City, Chicago, Philadelphia, and Detroit. Los Angeles's problem is probably due to three causes: extensive use of vehicles, considerable chemical industry, and topographic characteristics that help hold pollution in the basin.

With respect to world cities, for many years Mexico City has been regarded as the major city with the worst pollution. For example, the text *Environmental Science* by Nebel and Wright cites this conclusion in a special section called "Mexico City: Life in a Gas Chamber." The text *Environment* by Raven, Berg, and Johnson takes the same viewpoint. The international comparison is more difficult because there is no uniform data collection system. The best source of data is probably the Global Environmental Monitoring System (GEMS) of the United Nations. A report issued in the late 1980s by this program and the World Health Organization did not include data for Mexico City or the Soviet Union, but it did show that some of the other most polluted cities were Shenyang and Beijing, China, the former being the worst covered by the report; Calcutta and Delhi, India; Tehran, Iran; Santiago, Chile;

TABLE 3-A

Worst cities in the US for criteria pollutants, with special attention to comparisons between the South Coast [of California] Air Basin and the rest of the country

Contaminant and Federal standard	Three worst US cities compared to worst basin cities in terms of days exceeding standard or regulated annual or quarterly concentration	No. days exceeding standard (or other measure)	Three worst US cities compared to worst basin cities in terms of regulated short-term concentration value or other short-term concentration	Maximum concentration (for regulated period or other period indicated)
Ozone, O3 0.12 ppm (235 µg/m3) 1-Hr. Avg	1. *Glendora, CA 2. *Crestline, CA 3. *Redlands, CA Hesperia, CA Victorville, CA Clovis, CA	118 103 103 57 19 17	1. *Glendora, CA 2. *Crestline, CA 3. *Upland, CA Hesperia, CA Houston, TX Marietta, OH	0.30 ppm 0.28 ppm 0.28 ppm 0.23 ppm 0.22 ppm 0.21 ppm (8 hr)
Carbon Monoxide, CO 9.4 ppm (10 mg/m3) 8-Hr. Avg.	1. *Lynwood, CA 2. *Hawthorne,CA 3. Denver, CO Spokane, WA Phoenix, AZ El Paso, TX	31 7 7 6 5 3	1. *Lynwood, CA 2. Denver, CO 3. Phoenix, AZ Seattle, WA Spokane, WA Las Vegas, NV	18.8 ppm 15.8 ppm 15.4 ppm 12.9 ppm 12.1 ppm 12.0 ppm
Sulfur Dioxide, SO2 365 µg/m3 (0.14 ppm) 24-Hr. Avg.	1. Hawaii (volc.) 2. St. Clair Co. IL 3. Ohio Co. KY *S. Coast Basin	5 1 1 0	1. Hawaii (volc.) 2. St. Clair Co. IL 3. Ohio Co. KY *Hawthorne, CA	0.219 ppm 0.200 ppm 0.144 ppm 0.035 ppm
Nitrogen Dioxide NO2 0.053 ppm (100 µg/m3) Annual Mean	1. *Pomona, CA 2. *Burbank, CA 3. *Lynwood, CA Denver, CO Elizabeth, NJ New York, NY	0.0507 ppm 0.0501 ppm 0.0455 ppm 0.0405 ppm 0.0383 ppm 0.0368 ppm	1. *Los Angeles, CA 2. Sacramento, CA 3. *Pico Rivers, CA Dallas, TX Denver, CO Trona, CA	0.30 ppm (1 hr) 0.28 ppm (1 hr) 0.27 ppm (1 hr) 0.26 ppm (1 hr) 0.26 ppm (1 hr) 0.24 ppm (1 hr)
Particulates (PM10) 50 µg/m3 Annual Mean 150 µg/m3 24-Hr. Avg.	1. *Ontario, CA 2. Philadelphia,PA 3. Stev. Cnty, WA Bakersfield, CA Post Falls, ID Paul Spur, AZ	78.9 µg/m3 72.7 µg/m3 71.5 µg/m3 62.9 µg/m3 59.9 µg/m3 59.6 µg/m3	1. Spokane, WA 2. Ontario, CA Kennewick, WA Post Falls, ID Juneau, AK Anchorage, AK	803 µg/m3 (24 hr) 649 µg/m3 (24 hr) 596 µg/m3 (24 hr) 592 µg/m3 (24 hr) 586 µg/m3 (24 hr) 565 µg/m3 (24 hr)
Lead, Pb 1.5 µg/m3 Quarterly Average	1. Iron Co., MO 2. E. Helena, MT 3. Omaha, NB *S. Coast Basin	4 (qtrs) 4 (qtrs) 4 (qtrs) 0	1. Madison Co., IL 2. Philadelphia, PA 3. Iron Co., MO *Industry, CA	11.77 µg/m3(max) 10.04 µg/m3(max) 9.70 µg/m3(max) 0.48 µg/m3(max)

Note: This table is based on 1992 data published by EPA and California's South Coast Air Quality Management District. Locations within the South Coast Air Basin are denoted with an asterisk. For each pollutant, the first three cities listed are the three worst (having the most days exceeding the standard or the highest concentration) in the US If none of the three worst cities are in the Basin, the second group will be the highest in the Basin to show how they compare. If one or more of the three worst sites in the US are in the Basin, the second group are the additional highest locations outside the Basin.

Jakarta, Indonesia; and Milan, Italy. These estimates were based on sulfur dioxide, suspended particulates, and smoke.

4.   [This problem is intended for automobile enthusiasts.] The problem of combustion-generated nitrogen oxides has been described in the text. What changes are being investigated for future automobile and truck

engines that may decrease such emissions? (Hint: potential sources of information would include newspaper articles, popular magazines related to science or technology, automotive magazines, on-line searches or discussion groups, or discussions with people familiar with this technology.)

Nitrogen oxides in vehicle engines are produced because of the reaction between nitrogen and oxygen in the air at the elevated temperatures that take place in the combustion chamber. They are more of a problem with diesel engines, where the temperatures are higher, but they do occur with both gasoline and diesel engines. The two logical strategies for reducing such emissions would be to prevent the formation of the compounds in the first place, generally by reducing temperatures within the combustion chamber, or to remove them from the exhaust stream before it is released to the environment. For the latter approach, the only promising technology is the use of some kind of catalytic converter to free most of the nitrogen back up as N_2, normal atmospheric nitrogen. It is technically challenging to do this at the same time that other catalytic converter component are processing CO or unburned hydrocarbons. (Non-catalytic reduction of NO_x can be carried out in some other environments, but not in vehicle engines that cycle widely.)

For the reduction of combustion temperatures, various possibilities are in use, development, discussion (and sometimes in stages of derision), such as: exhaust gas recirculation (by injection into the cylinder at about the time of ignition), extra lean fuel mixtures (which result in lower temperatures, although slightly lean mixtures can increase NO_x by making more oxygen available to combine with the nitrogen); extra rich fuel mixtures, thus really keeping the oxygen away that might combine with nitrogen, although then you need some way to recover the energy value of the unburned hydrocarbons; use of catalysts on the wall of part of the combustion chamber, such as the face of the piston, to facilitate and control the combustion process and keep it at lower temperature; fuel modifications (again to control temperature of burning); careful design and control of the ignition and combustion process, so that no higher temperatures are created than those needed for efficiency; and even the injection of water itself to lower temperatures in the combustion chamber, as is practiced in many industrial boilers.

3.4 Brief Primer on the Exponential Function e^x

For Exercises 1–4, use your calculator or computer to determine sufficient values of the indicated function so that you can draw a reasonable sketch of its graph on the interval from -2 to 2. Pick a single common vertical scale for all the graphs that will be sufficient to cover enough of the corresponding y interval so that the differences among the graphs can be seen easily. (Feel free to use the computer to draw all the graphs as well if you know how.)

1. $y = e^x$
2. $y = e^{-x}$
3. $y = e^{x^2}$
4. $y = e^{-x^2}$

See Figure 3-B for the graphs corresponding to the first four exercises.

5. Carry out sufficient calculations to enable you to develop rough sketches of the graphs of $y = e^{-x^2}$ and $y = e^{-x^2/5}$ in a format that will enable you to compare them.

These graphs are shown in Figure 3-C.

6. Can you use the laws of exponents from algebra to simplify the expression: $e^2 \times e^3$? Check any result by evaluating the old and the new expression with your calculator or computer.

$$e^2 \times e^3 = e^{2+3} = e^5$$

$$7.3891 \times 20.0855 = 148.4132$$

7. Can you use the laws of exponents from algebra to simplify the expression: $e^2 + e^3$? Check any result by evaluating the old and the new expression with your calculator or computer.

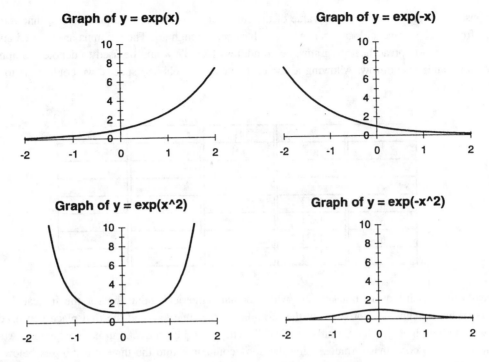

FIGURE 3-B
Graphs for Exercises 1-4 on a common scale

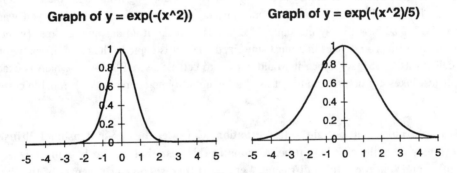

FIGURE 3-C
Graphs for Exercise 5 on a common scale

No; the laws of exponents do not provide any useful simplification of this expression.

3.5 One-Dimensional Diffusion

Several of the exercises in this and the subsequent section are ideally carried out using spreadsheet programs. I recommend that you try to program one or two of these yourself and check them against the results in this manual. This will let you anticipate the kinds of difficulties the students may have (although they are frequently more adept at spreadsheets than their teachers). The only tricky aspect of using spreadsheets for these and subsequentproblems is the distinction between relative and absolute addresses, so you may want to review this yourself and perhaps plan a demonstration for the students. The idea is explained below.

Suppose you want to calculate the value of the function $z = ax$ for some parameter a that you may want to change from time to time and for a range of x values, x_1 through x_4. The columns are lettered and the rows are numbered in most spreadsheet programs, so an address like B7 would be used to denote the number in the second column and seventh row. Allowing a few cells for text labels and spaces, we could start to set up this spreadsheet as follows:

	A	B	C	D
1	a=	[a]		
2				
3	x	z		
4	[x1]			
5	[x2]			
6	[x3]			
7	[x4]			

Here the entries in square brackets are where actual numerical inputs go for the indicated variable or parameters. What remains is to program in the formula for the function, and the first place you need this is in cell B4, where you might type the formula "=B1*A4". That would be correct, but not convenient. After all, you are going to want to copy that formula, using the copy command, into the three other boxes below it. (Think of the formula as some wild exponential mess that you want to type and debug only once.) The spreadsheet "copy" command tries to update logically the addresses of the input cells when you do this, so, when you copy the formula from cell B4 to cell B5, it revises the formula so as to choose all input cells according to the same relative position to the formula cell as they had in the cell you are copying from. Thus the copied formula would show up in cell B5 as "=B2*A5". The A5 is fine; in fact, that change is exactly what you wanted. But you really wanted the B1 address to stay the same. The key is to make it an absolute address by putting a dollar sign in front of any of its coordinates that must stay fixed when it is copied. Therefore, if you wrote the original formula in cell B4 as "=B$1*A4", then it would copy into cell B5 as "=B$1*A5", which is exactly what you would like. It just takes a little practice to get used to using absolute addresses with reliable correctness.

The following problems all pertain to this situation: Consider the one-dimensional diffusion problem as discussed previously in terms of a long tube. Assume in this case that a solid is being released into a liquid that fills the tube, and that the solid is diffusing. Let $x = 0$ correspond to the center of the tube and as usual use negative values to the left and positive values to the right. If a mass of 10 grams is released at the center of this tube, we are interested in knowing the concentration distribution along the tube as time increases. Assume that for the materials involved (the solid and the liquid), the diffusion constant $D = .05$ cm^2/sec.

1. Calculate the concentration at all of these locations: $x = -2, -1.5, -1.0, \ldots, +2$ at each of the following time values: $t = 1, 10, 100, 1000$. (Hint: it is strongly recommended that you use a spreadsheet program or other computer program to carry out these calculations easily. Furthermore, test a few of your calculations with a calculator to make sure you did not make a programming error.)

Figure 3-D shows tabular and graphical results for this and the next problem, calculated with a spreadsheet program.

2. Graph these results, preferably by computer, in the form of curves showing concentration as a function of location, a different curve for each time value. Try to get all the graphs on one set of axes.

See Figure 3-D.

	1	10	100	1000
-2	2.6E-08	0.53991	1.032883	0.391043
-1.5	0.000164	1.295176	1.127332	0.394479
-1	0.085004	2.419707	1.200039	0.396953
-0.5	3.614448	3.520653	1.245895	0.398444
0	12.61566	3.989423	1.261566	0.398942
0.5	3.614448	3.520653	1.245895	0.398444
1	0.085004	2.419707	1.200039	0.396953
1.5	0.000164	1.295176	1.127332	0.394479
2	2.6E-08	0.53991	1.032883	0.391043

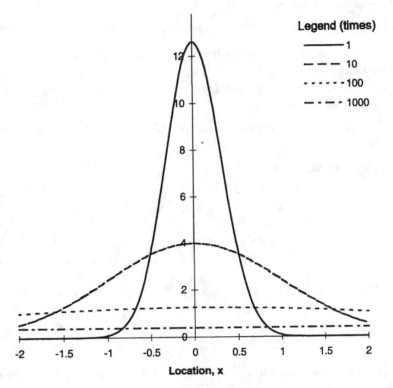

FIGURE 3-D
Calculational summary and graphs for Exercises 1 and 2

3. What are the correct units for the concentrations that you have calculated?

The concentrations are in grams per centimeter.

4. Explain in your own words why the graphs look reasonable (assuming that they do).

At earlier times the material is concentrated near the injection point, $x = 0$. As time goes on, it spreads out into a wider and less concentrated pattern. The patterns are all symmetrical, also as expected.

This next problem carries some of the previous themes further in asking the student to try out separately different logical attacks on a problem situation. This is really a theme of the entire book, namely, the need to constantly shift back and forth in applied problems between intuitive, physical reasoning, mathematical analysis of equations, and computational experiments. You can certainly give the students the option of using any approach that they are most comfortable with, but this problem, at least as it is written, is intended to require them to see how far they can go with each approach on a sequence of new situations. One way to use this problem is to let different students or groups present distinct approaches in class, and then discuss with the whole class which ones they feel most comfortable with and which ones they "would never think of." For weaker or more elementary students, who may think of a math course simply as a sequence of computational techniques to be learned, these kinds of discussions of problem solving are especially valuable. More advanced students will easily pick up these multiple lines of reasoning with not nearly so much processing and discussion being necessary, although often they make too little use of the physical intuition approach and even the numerical experiments.

5. This problem makes reference to your work on the previous problems. For each of the following situations, you are asked how the original answer to the diffusion situation would be modified if certain input values were changed. You must provide three distinct lines of reasoning to support your answer: first, an explanation based on your intuitive understanding of the physical diffusion process; second, an explanation based on the actual diffusion equation; third, an explanation based on recalculating your spreadsheet or program under the modified condition(s).

a) What should happen to all the entries if the mass M you start with is increased by a factor of two?

i)   All entries should double because twice as much mass can be thought of as two "copies" of the original amount of mass injected at the same time in the same place. Assuming they don't interact in any way to affect diffusion, the results should just superimpose on each other, yielding twice the concentration at every point. (This is the principle of superposition, which is discussed in chapter 6.)

ii)   Since M is just a multiplier on the diffusion equation, doubling it should double the concentration value for any point.

iii)   Recalculated spreadsheet values are as follows:

	1	10	100	1000
-2	5.2E-08	1.079819	2.065766	0.782085
-1.5	0.000328	2.590352	2.254665	0.788959
-1	0.170007	4.839414	2.400078	0.793905
-0.5	7.228896	7.041307	2.49179	0.796888
0	25.23133	7.978846	2.523133	0.797885
0.5	7.228896	7.041307	2.49179	0.796888
1	0.170007	4.839414	2.400078	0.793905
1.5	0.000328	2.590352	2.254665	0.788959
2	5.2E-08	1.079819	2.065766	0.782085

b) What should happen to the entry for $x = 0$, $t = 100$, if D is decreased by a factor of two?

i)   This is the injection point. Lower D means lower diffusion rate, so more material should stay around this point longer. (That is, it should not dissipate so fast.) Hence the entry should be larger.

ii)   For $x = 0$, the exponential term drops out. Multiplying D by $1/2$ should bring a factor of $1/\sqrt{2}$ out of the radical in the denominator, equivalent to multiplying the fraction by $\sqrt{2}$, or 1.414. Using the value in Figure 3-D for this location and time, the new value should be $1.261566 \times 1.414 = 1.7839$, although there will be some loss of precision by our approximation to the square root of two.

iii)   The recalculated spreadsheet is shown below with the value highlighted. The value is close to the one calculated previously, the difference being due to round-off errors.

	1	10	100	1000
-2	7.58E-17	0.103335	1.195934	0.542067
-1.5	3.02E-09	0.594651	1.424652	0.551637
-1	0.00081	2.075537	1.614342	0.558576
-0.5	1.464498	4.393913	1.740074	0.562781
0	17.84124	5.641896	1.784124	0.56419
0.5	1.464498	4.393913	1.740074	0.562781
1	0.00081	2.075537	1.614342	0.558576
1.5	3.02E-09	0.594651	1.424652	0.551637
2	7.58E-17	0.103335	1.195934	0.542067

c)   **What should happen to the entry for $x = 1$, $t = 100$, if D is increased by a factor of 10 (i.e., an order of magnitude) to a new value of 0.5? Will the same direction of change (i.e., increase or decrease) *always* be experienced at every x value (except perhaps $x = 0$) and at every t value (extending beyond those specific values in your table) whenever D is increased? Discuss why or why not.**

i)   Increasing D increases the diffusion rate, and so the material should dissipate faster from around the source. From the general shape of the graph for $t = 100$ in Figure 3-D, the level is already pretty uniform in the interval of interest, so faster dissipation would surely reduce the concentration on this interval and hence at the particular point $x = 1$. This need not happen at every point. For example, if you consider a point way far off to the right that has a negligible concentration at $t = 100$, increasing D should serve to get material to that point faster, so that the concentration at $t = 100$ could well be higher in this case.

ii)   Looking at the diffusion equation, we see that D always enters as part of the factor Dt. Therefore, increasing D by a factor of 10 has the same effect as would increasing t by a factor of 10 instead. This is equivalent to looking at the original process 10 times later, or at $t = 1000$. Thus the graph for $t = 1000$ in Figure 3-D also gives the concentration for the situation in this problem, illustrating a reduced C value at $x = 1$. (You might say that changing D simply changes the 'time scale' of the problem.)

iii)   The recalculated spreadsheet below bears out the point above; observe how the columns correspond to those from the original problem (Figure 3-D) but moved to the left.

	1	10	100	1000
-2	0.53991	1.032883	0.391043	0.125905
-1.5	1.295176	1.127332	0.394479	0.126015
-1	2.419707	1.200039	0.396953	0.126094
-0.5	3.520653	1.245895	0.398444	0.126141
0	3.989423	1.261566	0.398942	0.126157
0.5	3.520653	1.245895	0.398444	0.126141
1	2.419707	1.200039	0.396953	0.126094
1.5	1.295176	1.127332	0.394479	0.126015
2	0.53991	1.032883	0.391043	0.125905

6.   **The diffusion coefficient D shows up in two distinct places in the one-dimensional diffusion equation. The point of this question is for you to investigate what happens to the concentration value at a fixed location x and at a fixed time t as the value of the diffusion coefficient varies. Notice, for example, that as D gets small, its occurrence in the denominator of the first factor would tend to increase the size of the concentration. However, at the same time, when D gets small, it causes the exponent in the exponential term to become a very large negative number, thereby making that exponential factor get smaller. Thus these two effects tend to work against each other. Similarly, as D gets large, the opposite occurs. So the question is really how do these factors balance out. You may answer this question by using clear physical explanations and/or numerical experiments, or by using calculus (if you have studied this subject). Be precise in summarizing your conclusions.**

Keep in mind that we are working at a fixed location x and fixed time t that will be the same throughout the problem. For us they are constants. It is as though the diffusion process is going on and we come along at this one given appointed time and take one measurement of concentration at this predetermined fixed point, and then we look at that one value. How does this value depend on the diffusion constant D?

If D were practically 0, the source would hardly disperse at all in our finite time interval, and so we would measure a very low concentration value. As D increases, diffusion becomes more rapid, and we will start to measure higher values. But if we let D get really big, diffusion will be so fast that by the time we come to take our measurement, practically all the material would have diffused on by, and once again we would get a very small value. Therefore, if we try to graph our concentration value, at this fixed point and time, as a function of D, the

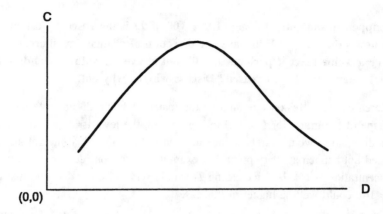

FIGURE 3-E
Initial estimate of the shape of the graph of the concentration C at a fixed point and time as a function of the diffusion constant D

graph should have the very general shape shown in Figure 3-E, although we can't yet be sure that there is only one maximum point.

So next we perform a numerical experiment. All points (except $x = 0$) should have the same qualitative behavior; the quantitative difference will be only a matter of scaling. Therefore we take the point $x = 1$, $t = 100$, with which we did some experiments earlier (Exercise 5), and we use the diffusion equation to graph the concentration C as a function of D. The results of such calculations are shown on the graph in Figure 3-F, which gives us a refinement of the cruder graph given earlier in Figure 3-E. Note that if you don't start with small enough values of D, you miss the peak and it looks as if the curve is always decreasing. The logarithmic scale helps us cover a wide range so as to avoid this problem. The figure shows the specific data points calculated and does not draw a smooth curve between them, simply to emphasize that we do not really know its shape there. It does look as if there is one maximum and that it is somewhere just to the left of 0.01. (One could refine the set of data points at which the calculations are performed in order to narrow down the uncertainty in the value of D that gives the peak.)

Now we apply calculus to see what we can learn about this situation. Since x and t are constants for this problem, C is just a function of the single variable D. One could investigate this curve using the derivative, but it looks messy because of where the D shows up in the radical and the denominator of the exponent. So, perhaps especially after some students have forged ahead and done this, it is good to point out that we can simplify the

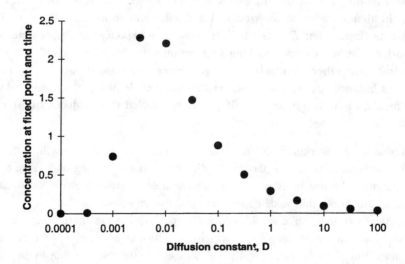

FIGURE 3-F
Results of numerical experiments on C vs. D, using a log scale for D to be sure to capture the full range of behavior

problem with a change of variable. Several choices are possible, but for here we will use

$$v = \frac{1}{\sqrt{D}}$$

so that the diffusion equation converts to

$$C = \frac{M}{\sqrt{4\pi t}} v e^{-\left(\frac{x^2}{4t}\right) v^2}$$
$$= A v e^{-Bv^2}$$

for the obvious positive constants A and B. The derivative is

$$C' = Av(-2Bv e^{-Bv^2}) + Ae^{-Bv^2} = A(1 - 2Bv^2)e^{-Bv^2}$$

so that the only positive critical point is at

$$v = \frac{1}{\sqrt{2B}}$$

and this is easily seen (first derivative test, by observation) to be a local maximum and absolute maximum for positive v. In terms of the original variables, this corresponds to

$$D = \frac{x^2}{2t}$$

For the numerical case used in the experiment, this yields the maximum at

$$D = \frac{x^2}{2t} = \frac{1^2}{2 \times 100} = \frac{1}{200} = 0.005$$

This is consistent with our earlier estimate that the point is just to the left of 0.01 (on the logarithmic scale).

3.6 Two-Dimensional Diffusion

The first four exercises below all pertain to the following situation: Suppose that 800 kg of soluble material are spilled into the center of a large shallow lake and gradually diffuse out into the lake. Because of the wind and current patterns, the diffusion constants are different in the east/west and north/south direction. The east/west diffusion constant is .3 m^2/min, and the north/south diffusion constant is .9 m^2/min. Use a spreadsheet or similar aid for the calculations. Let the x direction be east/west and the y be north/south, and assume the spill is at the origin. (Be sure to check a few answers with your calculator in case of programming errors, and be careful in using relative and absolute cell addresses if you use a spreadsheet program.)

1. **Find the concentration at the point $x = 15$ meters, $y = 20$ meters, and time $t = 10$ minutes. Be sure to label your answer with the correct units.**

$$1.32 \times 10^{-12} \text{ kg/m}^2.$$

2. **Consider the point $x = 2$ meters, $y = 2$ meters. Calculate and graph the concentrations at this point for $t = 10, 20, 30 \ldots 1000$ minutes.**

   This graph is shown in the top part of Figure 3-G for the t interval requested.

3. **Do the resulting values from Exercise 2 look like what you would expect from intuitive reasoning? Explain your answer.**

   No. You should expect to see the values start out near 0 for early times, and then decrease only after passing a peak. The problem is that the peak must have passed by the 10-minute mark, the earliest time for which Exercise 2

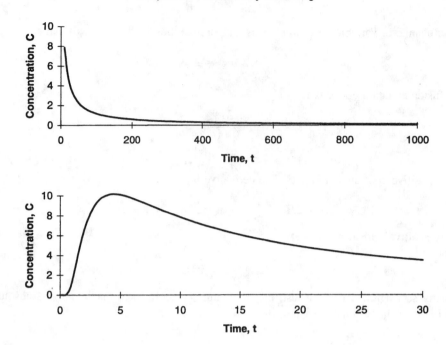

FIGURE 3-G
Calculations for Exercises 2 and 3

sought concentrations. The second graph in Figure 3-G expands the early part of the time interval and does indeed demonstrate the expected peaking behavior.

4.   Demonstrate by means of some concentration calculations of your own choosing and/or the corresponding graphs that in the situation under discussion, there really is a different rate of material distribution by diffusion in the x- and y-directions.

We will pick corresponding arbitrary but fixed "observation points" on each of the two axes, say, $(3, 0)$ and $(0, 3)$, and we will monitor the concentrations at these points as time increases. We should see a faster and more concentrated passing of the peak and higher concentrations at least past the peak point in the y-direction (namely, at the y-axis observation point) because of the larger diffusion constant in this direction. Figure 3-H shows these results, again obtained essentially painlessly by modifying the inputs on the spreadsheet developed for the earlier problems.

There are also other ways to do this problem. For example, one might compare the concentration distribution along the x-axis (east/west) to the concentration distribution along the y-axis (north/south) for a fixed time, say, $t = 10$. In this case you would see faster diffusion (and hence dissipation) in the y-direction because of the higher diffusion constant for that direction. This is illustrated in Figure 3-I. (Note: you might expect the peak to be smaller in the second graph, because of the faster diffusion. To check your or your student's grasp of the situation, why are the peaks at the exact same height?)

5.   Use techniques from basic algebra, including one of the laws of exponents, to rewrite the two-dimensional diffusion equation as a product of three factors. The first factor should be the total mass M, and the second two factors should correspond to terms that involve diffusion individually in the x- and y-directions, respectively. Furthermore, the second two factors should also have an identical form, except for whether their input values correspond to the x- or the y-direction.

$$C = M \left[\frac{1}{\sqrt{4\pi t D_1}} e^{\left(-\frac{x^2}{4D_1 t}\right)} \right] \left[\frac{1}{\sqrt{4\pi t D_2}} e^{\left(-\frac{y^2}{4D_2 t}\right)} \right]$$

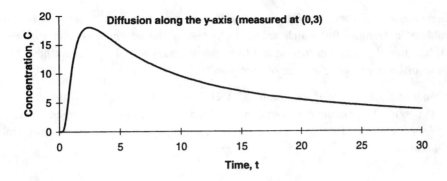

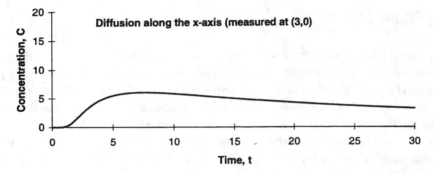

FIGURE 3-H

Comparison of diffusion in x and y directions (Exercise 4)

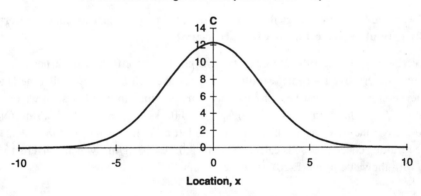

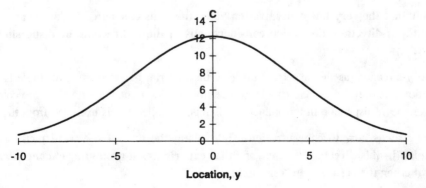

FIGURE 3-I

Alternative approach to Exercise 4

6. Using the new version of the two-dimensional diffusion equation you developed in connection with the previous problem, determine for which values of x and y the second two factors have their largest possible values. What does this location correspond to physically, and why does this mathematical observation correspond to an intuitively obvious physical observation?

Since the exponentials are "decaying exponentials," their largest values are at $x = 0$ and $y = 0$, respectively. Thus the origin is the point of maximum concentration. This is as expected since the highest concentration should always be at the point of injection of the source material, from which it constantly disperses away by diffusion into regions of lower concentration.

3.7 The Basic Plume Model

Note: It would be very useful in carrying out the plume calculations in these exercises for you to program the Gaussian plume equation into a spreadsheet program, another computer program, or a programmable calculator. This will eliminate the tedium of extensive hand calculation and will let you focus on using numerical experiments conveniently to answer some of the questions asked. In connection with use of this text, you may have received or have access to a special spreadsheet which already contains a program for the Gaussian plume model and that automatically incorporates the dispersion factors that you would otherwise have to calculate manually from Figures 3-16 and 3-17. If you elect to use this program, be sure to read the instructions on the top few lines. If you are developing your own program and would like an analytical way to represent the curves given in Figures 3-16 and 3-17, see Exercise 11 below prior to undertaking the earlier exercises, or, if it is within your capability, use a curve-fitting method to set up your own approximation to the curves in Figures 3-16 and 3-17.

1. Repeat the power plant example carried out in the text, using your spreadsheet or alternative program. (Hint: be especially careful to convert units when necessary.)

As discussed earlier, we will use the physical stack height for the effective stack height, a conservative assumption (perhaps too conservative for some real-life situations). We will also assume that the land is flat and that the receptor is at elevation 0. The spreadsheet provided uses metric units, in which system we obtain the following converted input values: $h = 106.68$ m, $u = 4.473$ m/s, $Q = .606$ kg/s, $x = y = 2,275.6$ m. (Note that it is not strictly necessary to convert the mass values. If the mass were left at 80 lb, the result of the calculation would be in units of lb/m^3.) The resulting concentration is calculated to be 9.02(E-19) kg/m^3, which converts to 5.62(E-20) lb/ft^3 and compares well with the value of 5.5(E-20) calculated manually in the text.

2. For the power plant example considered in the text and in the previous problem, answer the following additional questions:

a) Would you say that the very low concentration value that was calculated is due primarily to dispersion considerations in the y-direction, dispersion considerations in the z-direction, or dispersion considerations in the x-direction?

Clearly in the y-direction, as the dispersion distance the material has to travel over there is so much greater than in the z-direction (1.414 miles vs. 350 ft). (The dispersion factors in the two directions are comparable in size.) As discussed in Volume 1, dispersion in the x-direction is much smaller and is excluded from the model.

b) Changing your focus now to concentrations right along the axis of the wind passing over the stack, determine the limits (in this direction) of the zone that experiences a ground-level concentration of at least 1 ppm. Explain whether the results seem reasonable.

Using the conversion factor in the text discussion, a concentration value of 1 ppm is roughly equivalent to $1.1 \times 10 - 7$ lb/ft^3 for nitrogen oxides, which converts to 1.77(E-6) kg/m^3. By perusing the output of the spreadsheet, and narrowing in to find the boundaries of the region in question, we find such concentrations along the axis from about $x = 529$ m to $x = 906$ m. The results show that both near the stack (before much material has had a chance to disperse downward enough) and far from the stack (by which point much of the material has been lost in the y- and upward z-directions) the concentration is below this threshold, although in the small intermediate range indicated here, it does exceed 1 ppm. (Note: 1 ppm is just a hypothetical threshold used for convenience in this problem.)

3.   A coal-fired power plant in Ohio emits sulfur dioxide at the rate of 80 grams per second from a tall stack 120 meters above the ground. The average wind speed is 8 meters/sec, although the direction is variable. Consider the pollutant exposure of an individual living 1,000 meters from the plant. Assume that the individual's residence has an elevation of 35 meters above the ground level at the stack. Based on meteorological data for this site, you may use a stability class of C and assume that the wind blows in the direction of the residence 25% of the time.

a)   During periods when the wind is blowing directly towards the individual's residence, calculate the ground-level concentration of this pollutant at the residence.

The input parameters should all be self-explanatory. Note that $y = 0$ since the wind is blowing "directly" towards the residence. The program yields the concentration $C = 1.02(\text{E-04})$ g/m^3.

b)   What are the correct units for the concentration you have calculated?

As indicated, the units are grams per cubic meter.

c)   Given that the wind is blowing in this particular direction, does this location experience the maximum ground-level airborne concentration? Please explain your answer and illustrate it, if necessary, with additional calculations.

No it does not. The numerical calculations show a peak around $x = 1,000$; but if you refine that by generating a series of values at each meter in that vicinity (for example, by choosing an initial point of 900 and incrementing by 1 for some distance), the peak concentration appears to be around 1,175 meters. (One could narrow it down further, but this is sufficient to answer the question that was posed.) Alternatively, one could simply experiment with points on each side of 1,000 to find that there are higher concentration values at nearby but larger x values.

d)   Suppose that when the wind is not blowing directly towards the residence, the pollutant concentration there decreases to a negligible level. What then would be the long-term average concentration at the residence, taking into account periods when the wind is blowing towards it and when it is not?

It would be the weighted average of these two kinds of periods, that is:

$$\text{average} = .25(1.02 \times 10^{-4}) + .75(0) = 2.55 \times 10^{-5}.$$

You would probably not want to assign Exercise 4 unless you also assigned Exercise 3. Exercise 4 asks the student to obtain the same kind of result, but without being led along step by step. It requires the student to have internalized the concept of Exercise 3.

4.   A chemical plant emits ammonia into the air at the rate of 10 milligrams/sec from an open waste lagoon, which can be considered as a ground-level release point. The average wind speed at the site is 4 meters/sec. Assume that conditions at this site are generally quite overcast and that the wind over the lagoon blows directly at a school, located 2,000 meters away, 20% of the time (and that the rest of the time the wind has no impact on this receptor location).

a)   What stability class best describes the site conditions for this area? Please explain.

As per the note on Table 3-1, stability class D should be used for heavily overcast conditions.

b)   Calculate the long-term average ground-level concentration of ammonia in the air experienced at the location of the school.

When the wind is blowing at the receptor, the concentration from the spreadsheet is determined to be 1.59(E-04). Taking the weighted average of this with the zero concentration that occurs 80% of the time, the result is 20% of this value, or 3.18(E-05) mg/m^3.

5.   A nuclear power plant emits radioactive krypton at the rate of 350 microcuries/sec from a tall stack 175 meters above the ground. The average wind speed is 10 meters/sec, and the wind generally blows from the west, although there is some variation in the direction along an arc ranging from southwest to northwest. To provide a simplified description, a meteorologist reviews the historical wind data from the site and summarizes it by saying that 60% of the time it comes from due west, 20% of the time it comes from 30° north of due west, and 20% of the time it comes from 30° south of due west. The meteorologist also suggests that a stability class of D would be most representative of this site's conditions.

a)   Calculate the long-term average ground-level air concentration of krypton experienced by an individual living 1,500 meters due east of the stack. The individual's residence is on a slight rise at an elevation of 20 meters above the ground level at the plant itself.

The general layout and geometry of this problem are shown in Figure 3-J. The downwind distance when the wind comes from 30° off due west is found simply by using a 30°-60°-90° triangle, so that

$$\text{distance} = 1,500 \times \frac{\sqrt{3}}{2} = 1,299.$$

Because of symmetry, we can combine the two 30° cases into one case that is assumed to occur 40% of the time. Therefore we use the spreadsheet to find the concentrations in the two model cases, and then we take the weighted average of the results. The result is:

$$\text{average} = .6(3.80 \times 10^{-8}) + .4(2.01 \times 10^{-26}) = 2.28 \times 10^{-8} \mu Ci/m^3.$$

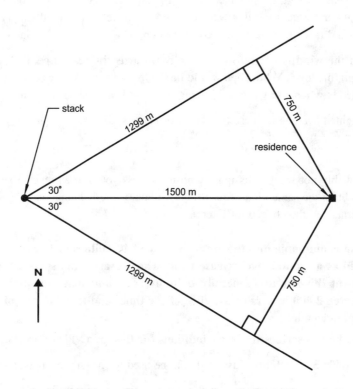

FIGURE 3-J
Plan view for Exercise 5

b) **In the calculation for part a, how would you describe the relative fraction of radiation received at the residence from times when the wind is blowing from 30° north or south of due west, as compared to that when the wind is blowing directly from due west?**

Negligible. As seen above, its value of 2.01×10^{-26} is many orders of magnitude smaller than the other value.

c) **Restricting your attention to the case when the wind is blowing directly from the west, find the downwind distance at which the concentration (at the same elevation as the residence) is largest. Explain what makes this location compare the way it does with the location of the residence.**

From the spreadsheet, we see that the value is around 9,000 meters, and then we narrow it down further to a value around 9,230 meters, or over 9 kilometers! The reason why this is so large is the result of the tall stack; it takes a long time for vertical diffusion to bring the maximal concentration down to this receptor elevation. This is why stacks are built tall, so that when the maximal concentration reaches ground level, it will already have dispersed considerably both upward and sideways.

6. **Analyze this situation and any assumptions made and decide whether the approach taken is logically valid. Give a *physically based intuitive explanation* for your answer, and then verify this with one or more *representative numerical calculations* using your spreadsheet or other program.**

You are applying for a permit to build a wood-fired boiler to generate heat and electricity for an industrial complex in central Maine, so you want to do some calculations to see how high the off-site pollutant concentrations might be, in particular at points of maximum off-site concentration, which will control the regulatory acceptability of the plant. The two predominant atmospheric stability classes experienced at your site are C and D. You limited your calculations for airborne concentrations of fine particulate matter to a variety of weather conditions within the D stability class because you believed that this was a "conservative approach" and would yield the highest concentrations, since a higher stability class means that the plume stays together longer and transports material farther from the source in higher concentrations.

The approach is not logically valid. While it is true that under D stability, all other factors being equal, the point of maximum ground-level concentration will be farther from the source, this does not mean that its actual value will be higher than the corresponding value under C stability. In fact, the maximum value under C stability will generally be higher, even though it will be closer to the source.

This is illustrated by the curves in Figure 3-K, which are based on a wind speed of 5 meters per second, which could correspond to either C or D stability conditions, depending on additional factors, and a unit source value ($Q = 1$) for comparison purposes. The stack height for these curves is 100 m. If the fence-line of the plant property (which is the nearest limit for the off-site concentration) is to the right of the point where the curves intersect, then

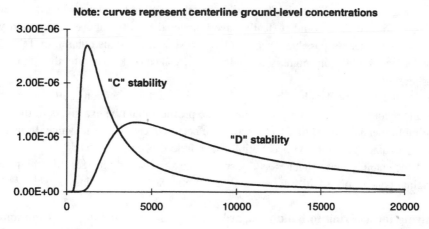

FIGURE 3-K

Comparison of stability class effects in Exercise 6

the D stability class would lead to the highest off-site concentrations. On the other hand, if the fence-line is closer to the stack than the point where the curves intersect, then the C stability class would lead to the highest off-site concentrations. Note that by varying wind speed and stack height, the location of this point of intersection could vary widely. [A good class exercise is to ask the students or student groups to predict how this point would move as the stack height is increased or decreased.]

7.   Analyze this situation and any assumptions made and decide whether the approach taken is logically valid. Give a *physically based intuitive explanation* for your answer, and then verify this with one or more *representative numerical calculations* using your spreadsheet or other program.

If a smokestack is emitting sulfur dioxide under a steady north wind of 6 mph, call the point of maximum ground-level concentration *A* and assume it is 1.3 miles downwind from the stack. Now suppose that the wind increases to 12 mph but that other atmospheric conditions remain the same. Since the same plume is just being moved along twice as fast, you assume that the maximum ground-level concentration will now be at a point *B* 2.6 miles downwind from the stack. (Two important cautions: First, be careful with your units. Second, be sure that any numerical example you give really satisfies every single aspect of the problem as described. For conditions or parameters not specified, you are free to choose your own values to illustrate your points as long as your values are reasonable.)

First, concerning the given information, you are not given the stack height or the stability class. The given wind speed could correspond to any stability class (i.e., any column in Table 3-1 in Volume 1, as well as overcast conditions yielding stability class D). In order to try to "control for" stability class, let us first try this problem under heavy overcast conditions, so that class D applies to both wind speeds.

The stack height is *not* a free parameter, a fact often overlooked by persons attempting this problem. It is constrained by the given information that the peak occurs 1.3 miles (= 2,092 meters) downwind. Therefore we experiment with the spreadsheet program so as to find an approximate stack height that yields a peak at this location. This value turns out to be 64 meters. This of course stays fixed for the second part of the problem, where we increase the wind speed to 12 mph (5.37 m/s) and see where the maximum occurs, in particular checking to see if it occurs at twice the distance from the stack, or 2.6 miles (= 4,184 meters). This second calculation yields what often looks like quite a startling result to some, namely, that the peak location has not moved at all!

After checking several times that they have not made a mistake, the better students usually return to the plume equation and explain why this is as expected.

> You can get the class to think about this issue by pursuing it as an in-class group project on the day they turn in their papers with solutions to the original exercise. First, you would probably want to review the approach that people took to the original problem and, in particular, what stability class assumptions they made to do the problem. Many probably will have just picked an arbitrary class and left it alone when they doubled the wind speed. This is not valid, and they would basically have blundered onto the above observation without a valid basis. Only certain transitions are possible since you have to stay in the same column of Table 3-1, except for the D-D case treated above. Once this point is cleared up, then you can ask them why they should have expected the D-D finding presented above.
>
> In particular, if the stability class hasn't changed, neither have the dispersion coefficients, which are simply functions of distance, not wind speed. Therefore the plume profile should be exactly as in the first case, except that the constant multiplier $1/u$ in the first factor will have been reduced to half its value. In fact, you can verify in your numerical calculations that the concentrations have indeed been cut in half.
>
> Naturally, if the stability class changes as a result of the wind-speed change, then things are less orderly still, and the original conclusion is still true that the logical approach put forth in the exercise is fallacious.

8.   You are applying for a permit to build a hazardous waste incinerator in your community. You already own your site, and you are converting an old power plant on that site, so the height and location of the stack are already fixed. In particular, the height of the stack is 120 feet, and it is located in the center of

your roughly circular piece of property, placing it **900 feet** from the property fence-line. The surrounding topography is approximately level.

Your principal pollutants of concern are "polycyclic aromatic hydrocarbons" (PAHs), which can be produced from incomplete combustion of organics. (PAHs can cause cancer.) These are often a component of the smoke or soot that can be produced during periods when something goes wrong with the combustion system.

a) If your most common daytime weather condition at the site is strong sunlight and a wind of 12 mph from the west, where would the point of maximum ground-level concentration be?

These conditions imply stability class C. The spreadsheet shows a maximum concentration at a distance of about 400 m, which you can narrow down to about 393 meters by generating a table of meter by meter concentrations. Converted back to feet, this would be about 1,289 feet due east of the stack, or 509 feet past the fence-line.

b) What would be the worst case weather conditions for off-site ground-level concentration? You may assume that average wind speeds less than 4 mph are too rare to require consideration.

The answer turns out to be B stability class with a 4 mph wind speed.

To obtain this solution, we narrow down the possibilities as follows, based on experience gained in previous problems, especially Exercise 7. For a given stability class, the concentration profile is fixed and only depends further on the wind speed by means of the multiplicative factor $1/u$. Therefore the highest concentrations occur when the wind speed is the lowest that can be experienced in that stability class. Therefore, based on Table 3-1 of the text and our restriction to wind speeds above 4 mph, we can narrow the comparison down to six specific cases: A (4 mph); B (4 mph); C (4.5 mph); D (4 mph—heavy overcast); E (4.5 mph); and F (4.5 mph). For the spreadsheet, we use a unit source strength.

Figure 3-L shows the concentration profiles for the first three cases. The last three cases have lower peaks farther to the right and have been omitted. A has the highest peak, but it is so close to the stack that by the time you get to the fence-line concentration, it has dissipated considerably. B also has a peak inside the fence-line, but it is still quite near that value just at the fence-line and immediately outside. The peak for C is outside the fence-line, but its value is still lower than the non-peak but fence-line value for B. Therefore the worst case conditions would be B stability, 4 mph wind. (For a unit source, this gives a concentration factor of 5.98(E-05).)

9. Based on the experience gained in the previous problems, write a concise summary of the relationship between atmospheric stability class, wind speed, stack height, point of maximum ground-level concentration, and actual concentration level at that point of maximum concentration.

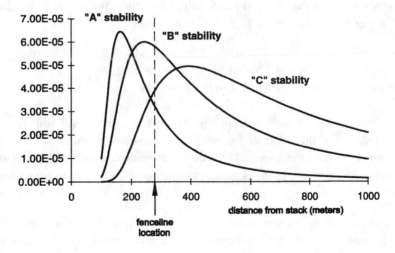

FIGURE 3-L
Comparison of concentration distributions under different stability conditions (Exercise 8b)

For a given stability class the point of maximum ground-level concentration will not depend any further on the wind speed (which has at least factored into the determination of stability class). It will depend on the stack height, for taller stacks will clearly result in points of maximum concentration that are farther from the stack. The value of that maximum concentration will indeed depend on the wind speed through the $1/u$ factor in the plume equation, which basically represents how much air is passing by in a fixed time to dilute the stack emissions. For higher stability classes, other factors being equal, the peak will be farther from the source and its actual value will be lower.

10.   One of the models that has historically been used to calculate the difference between physical stack height and effective stack height is called *Holland's equation*, which is given by

$$\Delta H = \frac{v_s d}{u} \left(1.5 + 2.68 \times 10^{-3} \times p \times \frac{T_s - T_a}{T_s} \times d \right)$$

where

$\Delta H =$ **the amount by which the effective stack height,**
exceeds the physical stack height (meters),

$v_s =$ **stack gas exit velocity (meters per second),**

$d =$ **inside stack diameter (meters),**

$u =$ **wind speed (meters per second),**

$p =$ **atmospheric pressure (millibars),**

$T_s =$ **stack gas temperature (degrees Kelvin),**

$T_a =$ **air temperature (degrees Kelvin),**

and 2.68×10^{-3} is a constant having units of $1/(\text{millibars} \times \text{meters})$. (There is actually a much more widely used formula due to Briggs, but it would introduce here some additional concepts that have little value for our specific purposes.)

You can see how complicated this calculation can be and that it introduces some additional physical parameters with which you are probably unfamiliar. Nevertheless, it will be useful to carry out one numerical example involving this equation, using realistic input values, to see how the effective stack height can vary under different conditions. In particular, a proposed source is to emit 72 grams per second of sulfur dioxide from a stack 30 meters high with an inside diameter of 1.5 meters. The effluent gases are emitted at a temperature of 250°F (394° Kelvin) with an exit velocity of 13 meters per second. The atmospheric pressure is 970 millibars, and the ambient air temperature is 68°F (293° Kelvin). Make up a table and graph to show the calculated difference between the physical stack height and the effective stack height H for a range of wind speeds from 1 meter per second up to 20 meters per second.

Figure 3-M shows the results of these calculations. Keep in mind that the wind speed corresponds to the speed of the wind at the top of the stack, and this speed might be quite a bit higher than at ground level or the often reported value for 10 meters elevation. For further discussion, see the next exercise.

11.   This is based on the results you have obtained in the previous problem. Using your physical intuition, explain why the trend shown in the calculated results in the previous problem is very reasonable. Use this same reasoning to describe under what general kinds of conditions you would expect the use of physical stack height instead of effective stack height in the Gaussian plume equation to lead to perhaps overly conservative calculational results.

As the wind speed increases, the momentum of the wind and the quantity of the air into which the stack feeds become much more dominant, thereby decreasing the potential for plume rise. Thus ΔH is lower for higher u values. Overly conservative results might thus be expected in cases of low wind speed, which, as we have seen in earlier exercises, may also lead to critical concentration values for determining the acceptability of off-site exposures.

wind (m/s)	$\Delta H(m)$	wind (m/s)	$\Delta H(m)$
1	48.74205	11	4.431096
2	24.37103	12	4.061838
3	16.24735	13	3.749389
4	12.18551	14	3.481575
5	9.74841	15	3.24947
6	8.123675	16	3.046378
7	6.96315	17	2.86718
8	6.092756	18	2.707892
9	5.415784	19	2.565371
10	4.874205	20	2.437103

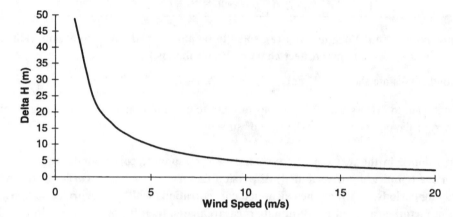

FIGURE 3-M
Calculations based on Holland's equation for stack height correction due to plume rise (Exercise 10)

12. With further reference to the model in Exercise 10, consider the following three input parameters: v_s, d, and T_a. From looking at the equation, what would be the effect on ΔH of increases in each of these three parameters, considered individually? For each case, explain on physical grounds why the direction of increase is what you would expect.

An increase in v_s causes an increase in ΔH since it enters as a multiplicative factor. This is expected since an increase in flue gas velocity should give the gases further momentum upward, thus causing them to require great time and vertical distance to be absorbed by the general wind movement.

An increase in d also causes an increase in ΔH since it enters both as a multiplicative factor in front and as a factor that increases part of the quantity in parentheses. This is also an expected effect since an increase in flue size, given that other factors like the velocity are the same, means that there is greater mass in the gas stream and hence greater vertical momentum.

An increase in ambient temperature T_a causes a decrease in ΔH since it decreases the numerator in part of the quantity in parentheses. This is expected since an increase in ambient temperature decreases the density contrast and buoyancy effect of the gases vis-á-vis the ambient air mass.

13. The use of the curves shown in Figures 3-16 and 3-17 would be much easier if you had an analytical representation for them (i.e., actual equations for the curves.) A number of different approximations, as well as refinements in the original curves for certain kinds of applications, have been developed over the years. One simple approximation is based on generating curves of the type found in Figures 3-16 and 3-17 in the form $\sigma = ax^b$, using different pairs of constants a and b to generate the six curves in each of the two figures.

Generate figures comparable to Figures 3-16 and 3-17 by means of this approach, using the following table of values for the constants:

	Horizontal dispersion, σ_y		Vertical dispersion, σ_z	
Stability class	a	b	a	b
A	.5169	.8689	.006948	1.638
B	.3585	.8748	.08336	1.0517
C	.2190	.8949	.11296	.9102
D	.1323	.8998	.2686	.6597
E	.0969	.9026	.3230	.5725
F	.0667	.8992	.2760	.5281

Use logarithmic axes, as in the original figures, so as to be able to make a comparison. (Hint: such axes are generally a standard option on spreadsheet chart or graph menus.)

The resulting curves are shown in Figures 3-N and 3-O.

You might want to discuss why these graphs turn out to be straight lines on a log scale graph, but it's not included as part of the question just for the sake of streamlining.

14. It was mentioned in the text that there should be a reasonable correspondence between the sampling times to which the (empirically determined) dispersion factors (σ_y and σ_z) correspond, and the length of the time-averaging periods for which the calculated concentrations will be interpreted to apply. For example, if the dispersion factors are based on 10-minute measurements, then if you use them in the plume equation to calculate a "long-term average concentration C," what C really corresponds to is a long-term average of 10-minute averages. (This is a subtle point, and you may need to think about it before addressing the following question.) Suppose you had two sets of curves of σ_y and σ_z values, one like Figures 3-16 and 3-17, based on 10-minute averages, and another new one based on one-day averages. How would the new set of

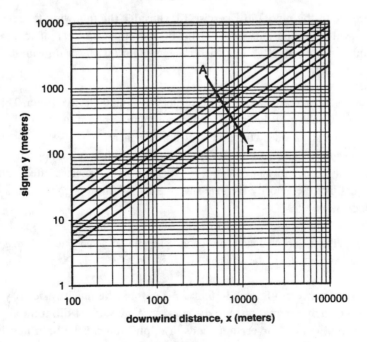

FIGURE 3-N
Graph of analytical approximations to σ_y dispersion factor curves

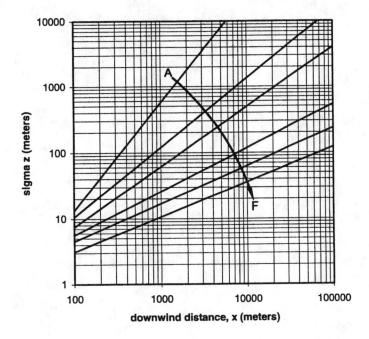

FIGURE 3-0

Graph of analytical approximations to σ_z dispersion factor curves

curves compare with the original set? If you have studied statistics, can you relate this to any concepts from that field?

The new curves should be lower, as longer-term measurements include more averaging of short-term variations, and hence they should exhibit less overall variability. In connection with statistics, this corresponds to what happen to sample means when the size of the samples increases. The variance (which measures variability) goes down.

3.8 General Comments and Guide to Further Information

(No exercises.)

4

Hazardous Materials Management

4.1 Background

The problems below ask you to find information about various classes of episodic events. You should be able to find such information fairly easily using your library's resources or via on-line searches. Newspaper searches, either computerized or via paper indexes, can be invaluable when you do not otherwise locate a database containing what you are looking for.

1. Aside from the hazardous materials incidents discussed in the text, identify one major incident in each of the categories below and provide a concise description of the incident, size of the affected population, materials involved, cause, and other relevant factors:
a)   rail transportation
b)   highway transportation
c)   oil well or pipeline
d)   fixed facility (e.g., factory, chemical plant, refinery)
e)   marine spill.

There is no shortage of such incidents, and they can be found by the various methods suggested. Several Federal agencies (e.g., EPA, DOT, USCG, NOAA) do maintain incident databases, and the students may come across these while doing on-line searches. But in the end, newspapers, as long as you have a good search tool, are just about as good for information on specific incidents. Some typical incidents in the various categories are given in Table 4-A below. This is not at all complete and is intended only to give some "yardstick" by which to judge the incidents submitted by the students. In fact, one might want to assign extra credit for the student finding the most severe incident in each category, as this may stimulate more thorough investigation.

TABLE 4-A
Examples of accidents and incidents involving hazardous materials

DATE	LOCATION	CHEMICAL	MODE	COMMENTS
4-4-97	Rostraver, PA	ammonium nitrate	highway	26 tons of fertilizer material spilled from an overturned trailer at the same time that diesel fuel also spilled; combined the mixture is explosive; closed I-70
12-96	Lake Zurich, IL	acid and caustic	highway	truck overturned with 11 portable tanks, some of acid and some of caustic, which react violently; some leakage; overnight evacuation of area
4-96	Alberton, MT	chlorine	train	derailment; multiple chemical leaks; toxic cloud was 8 miles long and 4 miles wide; town evacuated for 3 weeks; 350 injured; 1 fatality

DATE	LOCATION	CHEMICAL	MODE	COMMENTS
3-96	Weyauwega, WI	propane	train	5 of 14 propane cars engulfed in flames; town evacuated for almost 3 weeks
2-96	Cajon Summit, CA	butyl acrylate	train	derailment of 44 cars and fire involving some chemicals; no release from tank car of most concern; Interstate 15 was closed for a day because of danger
2-96	Leadville, CO	sulfuric acid	train	derailment; acid gas cloud; evacuation of nearby businesses and residents
96	Frenchtown, NJ	helium, oxygen, acetylene, propane	plant	plant fire; allowed to burn out because water could not be used to fight fire; half-mile evacuation zone
10-25-94	Usinsk, Russia	oil	pipeline rupture	23 million gallons released
8-93	Tampa Bay, FL	fuel oil	marine	collision of a freighter and 2 barges; leaks from two; 280,000 gal spilled
7-93	Anacortes, WA	ammonia	plant	fire in dock facility impinged on anhydrous ammonia tank, half-mile evacuation zone, 1500 lbs released, facility collapsed in water
12-26-92	Titusville, FL	phosphorous trichloride	train	derailment;400 evacuated within a half-mile zone
12-3-92	Aegean Sea tanker, off Spain	oil	marine spill	21.9 million gallons released in harbor area
9-30-92	Reading, MA	gasoline	highway	collision of tank truck and car; 11,000 gal spilled, closing I-93
7-92	Duluth, MN	benzene	train	60,000 evacuated
4-22-92	Guadalajara, Mexico	hexane	sewer system	multiple explosions leveled more than 20 city blocks; 170 deaths, 500 injuries; material leaked into sewer system and spread under part of city
11-91	Shepardsville, KY	methyl diphenyl diisocyanate; explosives	train	derailment; chemical similar to MIC released in Bhopal; concerns over BLEVE; 1000 evacuated
6-16-91	Dunsmir, CA	metam-sodium (pesticide)	train	19,000 gallons into Sacramento River; 5,000 evacuated
5-30-90	Freeport, TX	various	plant	chemical plant; 2,000 evacuated
10-89	Wellston, OH	ammonia	plant	release from pizza plant; entire town of 7500 evacuated; about 50 injuries
7-22-89	Freeland	acrylic acid	train	3,000 evacuated
1-6-89	Simi Valley, CA	chlorine	plant	textile plant (chlorine for bleaching); 12,000 evacuated
89	Helena, MT	various	train	runaway train; derailment; 3,500 evacuated
9-88	Los Angeles area, CA	chlorine, trichlorine-triazine-trione	plant	swimming pool pellets, fire, 28,000 evacuated
6-88	Springfield and Chicopee, MA	chlorine	plant	swimming pool pellets, rain in window, fire; 25,000 evacuated
6-88	Crofton, KY	chemical fire	train	36 cars derailed; 15,000 evacuated
5-88	Lincoln Heights, CA	fire	plant	metal plating plant; 11,000 evacuated
2-88	northwest OH	toluene	pipeline rupture	84,000 gallons; 5,000 evacuated

DATE	LOCATION	CHEMICAL	MODE	COMMENTS
10-87	Texas City	hydrogen flouride, isobutane	plant	storage vessel failure; 1041 injuries to members of public, 5,800 evacuated
5-87	Pittsburgh, PA	phosphorous oxychloride	train	derailment; evacuation of surrounding residents from incident and then during clean-up
6-3-79	Ixtoc No. 1 Well, Mexico	oil, gas	well	offshore well blowout; 140 million gallons released
3-16-78	Amoco Cadiz, off France	oil	marine spill	68.7 million gallons released

2.   For either your home state or the state in which your institution is located, find five examples of hazardous materials incidents that have occurred in the last two years and provide brief summary information on each.

Newspaper searches easily lead to many examples, which may vary widely in severity. Other sources may include state logs of incidents for which hazmat response teams have been summoned, and, in fact, these may even be on-line. A short portion of the 1996 on-line log for Massachusetts follows:

1/6 Haz-mat Belchertown, MA. 28 South Main St., 200 gallons heating oil spill. 17:56 hrs.

1/11 Haz-mat Boston, MA. Box 235 - Brookline Ave. and the Fenway, natural gas leak. 21:12 hrs.

1/11 Update Boston, MA. 2nd alarm haz-mat, gas leak - evacuations in progress. Special call Rescue 1. 21:33 hours.

1/11 Further update Boston, MA. Special call Ladder 17 with tunnel rescue trk. and meters, masks. 22:28 hrs.

1/15 Level 3 haz-mat, Boston, MA. 155 Southampton St., Russer Foods, Ammonia leak. 07:16 hrs.

1/15 Haz-mat Lynn, MA. Mt. Pleasant and Essex Sts. Major gas leak, evacuations in progress. 13:53 hrs.

1/17 Haz-mat Fall River, MA. Whipple & Manton Sts. Natural gas leak with evacuations. 12:51 hrs.

1/22 Haz-mat/ Motor vehicle accident, Quincy, MA. Box 1226, Burgin Pkwy., and Dimmock St. Oil truck rollover and spill. 09:32 hrs.

1/22 Update haz-mat Quincy, MA Box 1226. Oil truck rollover w/ 1900 gallons fuel oil leaking. 09:34 hrs.

1/24 Working fire Winchendon, MA. 202 Spruce St., Magnesium fire in bldg. 11:29 hrs.

1/24 Update working fire Winchendon, MA. Fire contained to magnesium bin, being removed. 11:36 hrs.

2/12 Working fire Fall River, MA. Currant Rd. at Molten Metals, commercial bldg. 21:50 hrs.

2/13 MVA/ Haz-mat Chelsea, MA. Box 63, Broadway and Williams St. Heating oil truck, spill of 500 gallons. 04:25 hrs.

2/13 Haz-mat Lynn, MA on Lynnway (old Norelco plant) gas main break, with evacuations. 15:13 hrs.

2/14 Haz-mat Wellesley, MA. Rte. 135 near Wellesley College. Limited response for Team C. 14:27 hrs.

2/28 Haz-mat Wakefield, MA. Vocational high school, Hemlock Rd. Mass casualty incident, CO problem with evacuations. 11:34 hrs.

2/28 Update haz-mat Wakefield, MA. Approx. 30 persons transported due to CO. 11:42 hrs.

3/7 Haz-mat Montgomery, MA. Russell Road, District 11 team activated. Propane truck off the road. 17:40 hrs.

3/7 Haz-mat Worcester, MA. 1200 Millbury Street. Plow struck Propane tank. 19:08 hours.

3/7 Update haz-mat Worcester, MA. 1200 Millbury St., 1200 lb. Propane truck, defensive mode. 19:16 hours.

3/11 Haz-mat Chelsea, MA. 11 Broadway, tanker (ship) docked at Tobin Bridge. 12:12 hrs.

3/11 Update haz-mat Chelsea, MA. Large volume spill, Chelsea Fire Dept., Coast Guard on scene. 12:14 hrs.

3/11 Haz-mat Somerset, MA. Rte.6 walk-in-clinic phenol spill, Dist. 1 team activated. 13:19 hrs.

4.2 Hazardous Materials Handling Practices and Potential Accidents

(No exercises.)

4.3 Physical Principles and Background

4.3.1 Basic Material from Physics and Chemistry

For some of the exercises below, you will need to consult some reference material of your own choosing, such as in your library or via computer. Sometimes even an ordinary dictionary or encyclopedia can be the most convenient source for data on elements or chemical compounds, or for conversion factors. The material needed to help answer these questions should not be difficult to find, and this initial practice in finding it will help guide you to sources that will be of further use later in this chapter.

1.   Find the molecular weights of benzene, sulfuric acid, and ammonia. (Reread the instructions at the top if you're puzzled on how to start.)

The molecular formulas for all of these can be found in a comprehensive dictionary, as well as on-line and in science books. The necessary atomic weights are all available in Table 4-2 in Volume 1.

Benzene is C_6H_6, so its M.W. $= 6(12.011) + 6(1.001) = 78.07$.

Sulfuric acid is H_2SO_4, so its M.W. $= 2(1.001) + 1(32.07) + 4(15.9994) = 98.07$.

Ammonia is NH_3, so its M.W. $= 1(14.001) + 3(1.001) = 17.00$.

2.   The atomic weight of carbon (and many other elements) has often been revised in the scientific literature. Doesn't this seem strange for such a common element? What do you think is the dominant reason why such values are revised from time to time?

The atomic weight depends on an estimate of the relative abundance of the different isotopes on the Earth. (Atomic weight is usually defined in terms of terrestrial values, not cosmic values.) But there are many different carbon reservoirs on earth that have different isotopic composition because of their age and the way they were formed (e.g., plant and animal cells, atmosphere, coral reefs, limestone formations, coal beds, oil and gas fields, dissolved CO_2 in the ocean and other bodies of water, etc.). Since our knowledge of these is constantly developing, it is not surprising that the estimated values keep changing slightly.

3.   Research the famous "Hindenberg disaster" and relate it to some of the material in this section.

The Hindenberg was a zeppelin, similar in principle to a modern-day blimp except that unlike a blimp, which forms its shape like a balloon as the result of the internal gas pressure, a zeppelin has a skeletal structure. Zeppelins were used extensively in World War I, and also maintained a thriving transportation business, including regular and quite luxurious transatlantic crossings, early in the twentieth century, before commercial flights by airplanes reached competitiveness. The Hindenburg was a "giant zeppelin," built by the Zeppelin Company in Germany and originally intended to be filled with helium, a light and non-flammable gas usually used in these airships. But because the US was worried about the military threat from such airships in the tense period leading up to World War II, the "Helium Act" was passed, making it impossible for the Germans to get the helium they needed for the Hindenburg. (The US had the only known natural helium reserves.) Therefore hydrogen was used, a lighter but highly flammable gas. As the Hindenburg was approaching its mooring point in Lakehurst, New Jersey, on May 6, 1937, it caught fire and fell the final 200 feet to the ground as a blazing mass. Thirty-seven crew and passengers were killed, but others were able to run from the gondola when it hit the ground. The cause of the fire was never determined for certain, and scientifically interesting new theories continue to be investigated. (In fact, one recent theory suggests that the hydrogen was only a minor contributor to the fire.) Ignition could have resulted from the discharge of a static electricity charge on the airship as it got close to the mooring point. (This is of great concern in working with flammable materials, as their movement through pipelines and between containers can generate a static charge that provides a potential ignition source upon discharge, just as you can generate a spark when you touch something after walking across a carpeted room.)

Despite its hazards, hydrogen is also very attractive as fuel. It is used in many rockets. But of particular interest is its long-term potential to replace fossil fuels in such applications as automobiles and other vehicles. It is widely available in water ($= H_2O$), although it would take energy from some source (e.g., an electrical power plant, perhaps someday powered by hydrogen fusion, as on the Sun) to separate it, and when you burn it with oxygen, the combustion product is just water again. These concepts are sometimes called the "hydrogen economy."

4.3.1.2 Atoms and Molecules

1. Consider the following five common elements: gold, silver, iron, lead, and uranium. Find their densities or specific gravities and place the elements in order from the lightest (i.e., the least dense) to the heaviest.

From lightest to heaviest, with the specific gravity value in parentheses: iron (7.9), silver (10.5), lead (11.3), uranium (18.7), gold (19.3). These values may be surprising. Remember that water has value 1 on this scale. The high value for gold makes it practically impossible to avoid detection of an alloy or other plated metal intended to pass as gold because the latter would not have the proper density. The high value for uranium is the reason why "depleted" uranium (uranium in which the fissile isotope U^{235} has been largely removed for other uses) is commonly used in aircraft nose cones to provide ballast and also used in armor-piercing artillery shells because its high mass and its deformation properties cause it to generate such intense heat instantly upon impact that it can melt its way through armor.

2. Sometimes the line between different states of matter is not so clear. For example, some people think that ordinary window glass is a solid, while others think it is a liquid. (Yes, it says "liquid." After all, if you look at very old windows, you might even see horizontal lines in them where they appear to have "slumped" over the years.) Investigate this issue with a science text or teacher, and summarize your conclusions.

There have been heated arguments over the years about how to classify glass. Proponents of the liquid classification argue that it is a "supercooled liquid." This means that as it cools down from the molten state present during manufacture, it gets so thick and "viscous" that even when it reaches the usual solidification temperature (= "freezing point"), the molecules are not mobile enough to reorganize themselves into the crystalline structure that would normally characterize the material's solid state (and that thermodynamic considerations would favor). Then it continues to be cooled to ambient temperature without changing the arrangement of the molecules into anything other than their usual configuration from the liquid state. Window deformation, as mentioned in the problem, and thicker glass at the bottoms in many old church stained-glass windows, have been cited to support this interpretation.

However, the prevailing scientific position today is that glass is indeed a solid. It is called an "amorphous solid" to recognize the non-crystalline or poorly ordered arrangement of the molecules. The modes of deformation studied under controlled laboratory conditions do not follow viscous fluid models, and the chemical bonds among the molecules are distinct from those present in most liquids. Window deformation is explained in terms of earlier manufacturing irregularities and other modes of deformation (like sagging beams and bending posts), and thick glass at the bottoms can also be traced to manufacturing processes, the intuitive preference of glaziers to use thicker glass at the bottom, and biased data samples.

Extensive discussion of this issue and supporting references can be found in on-line discussion forums by searching on several of the key words in the above paragraphs.

3. What is the atmospheric pressure in Denver compared to that in New York? How long would you have to boil an egg in Denver to reach the same result as with a given boiling time in New York? (Hint: you will have to be a clever investigator, chef, or scientist to answer this latter question. It is a well-known issue.)

Denver is sometimes called the "mile-high city" because its elevation is about a mile above sea level. The atmospheric pressure there is about 12.3 psi ("pounds per square inch"), compared to about 14.7 psi at sea level, so this is a reduction of about one-sixth. The result is that the boiling point of water is reduced to about 203°F from its sea level value of 212°F. (Remember, this is because when you heat water anywhere to 203°F, its vapor pressure reaches 12.3 psi; but in Denver this is actually enough to let it go ahead and boil, which also prevents its temperature from rising higher.) Books on science, cooking, and camping all often touch on the problems of cooking under high altitude conditions. The *Joy of Cooking* suggests adding 10% to the boiling time for each 1000′ of increased elevation, at least for a range of larger or whole vegetables listed there in which heat transfer through the solid would take some time, similar to the case of a whole egg. This would correspond to a 50% increase for Denver. But there is no quantitative discussion of soft-boiled eggs in particular.

Determined to settle this issue definitively and frustrated by e-mail correspondence with chefs at restaurants in Denver with on-line computer listings, the author attempted to conduct a scientific experiment in his kitchen, armed with about two dozen similar-looking eggs and a thermometer to simulate the temperature of boiling water in Denver. The egg supply was depleted before he felt he had established either an adequate control or adequate repeatability, a reminder perhaps of why he never elected to go into laboratory science.

Further searching led him to the book *A Newcomer's Guide to Colorado*, which, in its section on egg cooking, mentions a small device that one can put in a pot of water with eggs that one desires to cook to a certain "doneness." The changes in color within this device signal when to remove the eggs from the water. Such a synthetic egg was quickly obtained from a local kitchen store. At last, a controllable experiment, at least from the standpoint of the egg! However, maintaining the water temperature at 203°F, the boiling point in Denver, is not too easy, so there was still a problem with the scientific plan.

Therefore, growing only firmer in determination to settle this matter, the author and his somewhat skeptical wife decided to climb to an altitude of one mile in the nearby White Mountains of New Hampshire, armed with appropriate scientific and cooking equipment. (This would also allow the investigation of Exercise 4, below, which had become even more controversial at gatherings of family and friends.) Perched just below the summit of Mount Madison, in a sheltered location, the synthetic egg was "boiled," and records were kept of the time required to reach each of the "degree of doneness" marks. It was also confirmed that the water boiled at 203°F. By comparing these results to those from a comparable experiment in the Boston area (sea level), it was found that to boil an egg at the altitude of Denver only requires approximately 10% longer than in Boston or New York.

4.   If you had two identical stoves and two identical pots of water on them in both Denver and New York, which one would come to a boil faster? Explain your reasoning.

The straightforward answer is that the one in Denver should boil first. Since heat is being transferred into the two pots at the same rate, their temperatures should rise at the same rate; and at the moment when both are at about 203°F, the one in Denver should start to boil. The one in New York would have to be heated longer to reach its boiling point of 212°F. I would consider this to be a perfectly acceptable and correct answer based on the scientific principles presented in the text.

But there may be a catch, since the author's experience in camping in the mountains suggested that it can take "forever" to boil a pot of water there, although this is based on situations in which there had been no careful control for ambient temperature or relative humidity. In any case, assuming ambient temperature and relative humidity are the same, there is still some variation in the two situations, deriving in part from the fact that the ambient pressure to which a liquid is exposed has some effect on the *rate* of evaporation. This effect is not usually significant, and, in fact, it is not even included in the evaporation modeling in Chapter 7. But in considering two cases whose only difference is an unusually large difference in ambient pressure, the effect may be important. What happens is that even though the evaporation process tries to achieve the same vapor pressure equilibrium point in both cases, in the high altitude case, the lower ambient pressure helps the molecules that are newly arrived in the vapor space to move away from the liquid farther with fewer of the kinds of collisions that might redirect them back down through the surface and into the liquid form again. Since we are not working in a sealed container (as had been illustrated in Volume 1 in Figure 4-4), the gas-to-liquid part of the equilibration process is reduced while the liquid-to-gas part stays the same, so evaporation actually takes place faster. But faster evaporation uses up more of the heat supply within the liquid, robbing it of heat that would otherwise raise the temperature, so the pot on the stove in Denver should show a slower temperature rise. But now we have a slower temperature rise and a lower endpoint to be reached, and it is not clear how the two factors balance out at this particular altitude. (There are also other effects that would work in the opposite direction, such as the fact that the evolving lower mass of water in the pot in Denver takes less heat then to raise its temperature further, so this is indeed an interesting situation.) It is worth pointing out that several cookbooks that discuss high altitude recipe modifications do indeed recommend increasing water content in order to compensate for more rapid losses during both boiling and baking.

As noted in the solution to Exercise 3, a hike up Mount Madison in New Hampshire enabled the author to collect some experimental evidence on this issue. The initial plan was to conduct the experiment with a small camping stove heating a pot containing one canteenful of water and having a loose top. But control of flame temperature and heat input is critical, especially at an altitude where the oxygen content of the air is less. Therefore a thick, round steel plate was fabricated to sit on the stove under the pot, and a hole was drilled into it from the edge to receive a thermocouple probe. This made it possible to monitor and control the plate temperature (by varying the fuel supply to the stove), which is the direct source of heat for the pot. The experiment was run in August, 1997, on a day when the temperature was 59°F near the top of the mountain. The water was 55°F at the start, and it took 31 minutes to reach a slight boil (the best that could be achieved). (The steel plate was controlled to 630±°F.) Then, when a day of the same temperature (and other similar weather conditions) was experienced in the Boston area, the experiment was repeated at that elevation. In this case, the water boiled in 26 minutes, even though the boiling point was higher by 9°F. While perhaps not a perfectly controlled experiment, it does illustrate that this issue may be more complex than one originally might have thought.

The issue pursued in Exercises 3 and 4 could form the basis for a more detailed investigation by a student with a strong scientific bent, such as a physics major looking for an undergraduate thesis project. First, the conduct of truly well-controlled experiments would be a challenge. Second, the scientific basis for the synthetic egg could be interesting—is the cooking of an egg dependent on end-point temperature or some integral of the time and temperature profile? (The designer or manufacturer of the device could be worth talking to.) Third, further factors could be included in the mathematical modeling—such as the variation of heat of vaporization with temperature, or the diffusivity of water vapor in air as a function of pressure. (*The Chemical Engineers' Handbook* is a good source of data and references on such issues.) At almost the opposite extreme, one could also investigate the underlying principles embedded in this problem by considering limiting cases for the geometry of the pot, the amount of water, the circulation in the area over the liquid, the degree of mixing, and the heat supply. Such limiting cases can be used to show that either pot could boil first, depending on such other factors. On another front, students may be interested in reading a very understandable article by Juan Negret in *The Physics Teacher* ["Boiling Water and the Height of Mountains," May, 1986, pp. 290–292], which reviews basic theoretical and experimental issues associated with historical strategies of using the boiling point to estimate the height of mountains.

5. **Return to the potential accident discussed in the text, where liquid propane is poured out on the ground and boils off rapidly. Even though it boils at ambient temperature, it still takes a fixed amount of heat (the heat of vaporization) to convert each gram to gaseous form. Thus it would take a huge amount of heat for a whole tank-truck load to boil off. Where would this heat come from? Explain how you would envision the scenario.**

The heat would come from the ground, just as when you put ice cream in a bowl, the bowl gets cold because its internal heat content is being transferred in part to the melting ice cream. Thus the ground beneath the spill would get colder and colder. As this happens, there is less heat for the ground to transfer into the pool (just as if you put ice cream in a very cold bowl, it won't melt as fast), and the boiling process may slow down and essentially stop. Naturally there would also be some heat transfer from the ambient air; but due to its much lower density, air has a much lower heat capacity than the ground. Therefore it is not as productive a reservoir of heat for boiling the propane. The issue of heat transfer from the ground to the pool is modeled in some detail in Chapter 7.

6. **Aside from propane, find another relatively common chemical that is used in liquid form but that boils at ambient temperature.**

The refrigerant liquid (e.g., freon or modern equivalents) in your refrigerator or air conditioner has this property, which is exactly what makes it work. Its evaporation from the liquid into the vapor form extracts heat from the air (through the thin pipes that carry the refrigerant), thereby cooling this air, which is circulated for its cooling effect. The compressor then converts the material back to liquid, with the resulting heat being vented outside the refrigerator or outside the house or vehicle with the air conditioner, and the cycle is repeated. Anhydrous ammonia is used for industrial refrigeration processes in the same way, and you may have come across hazmat incidents involving its leakage from skating rinks, food plants, or other facilities that use refrigeration. Liquid butane is used in some cigarette lighters. Liquefied natural gas (LNG) is the liquid form of natural gas and is convenient for storage and transportation in certain circumstances, such as in transoceanic commerce, when pipelines are not available.

7. Can you identify the common substance alluded to in the text that does not necessarily expand when heated (or contract when cooled)? What is the significance of this behavior?

Water, in the limited temperature range from 0°C to 4°C (that is, just above its freezing point of 0°C) actually expands when it is cooled. The same is true as it changes state from liquid to ice just at the freezing point, 0°C. Above 4°C it behaves more normally and starts to expand again when heated. In other words, liquid water reaches its densest form at 4°C.

The expansion of water as you move from 4°C down to 0°C can be thought of as the beginning of the process by which the water molecules align themselves in the general framework that will eventually form a solid ice crystal at the freezing point. This expansion of water upon freezing, a very unusual kind of behavior, is one of the reasons that lakes do not generally freeze solid, a fact that facilitates the winter survival of fish and other forms of life and hence is key to survival of many life forms in cold environments.

The freezing process goes as follows. As the surface cools, at least in the realm above 4°C, the colder water, being more dense, sinks. Over time, as the lake water continues to be cooled at the surface by sub-freezing temperatures, the vertical temperature profile in the lake should be such that any 4°C water should collect at the bottom, with a gradual decrease in temperature to freezing and decrease in density (i.e., becoming more buoyant) as you move up towards the surface. So the densities in the water column should now be in equilibrium, and as there is further cooling from above, there should be no tendency to sink or otherwise mix. Ice then forms at the surface, and it is even less dense than the liquid water, so it stays up on top. Now any further cooling or freezing that is going to take place is going to have to be as a result of heat conduction from the top layer of water up through the ice. (Remember, you have to remove 80 calories per gram to change liquid water at 0°C to solid ice still at 0°C.) As heat is removed, that upper layer of water stays there, freezes, and the ice gets thicker. Heat transport by conduction alone is a slow process, especially as the ice gets thicker and the temperature gradient decreases. It would take the movement of a tremendous amount of heat by this slow process to bring the water in the column, the lower parts of which are several degrees above freezing, down to the freezing point and, in addition, remove the heat necessary for the freezing process itself. This is why the ice in the coldest parts of the US still rarely gets thicker than a few feet.

Since the entire lake over its full depth would have to be brought down to 4°C before the cycling ends and the surface layer could stay there and be taken down in temperature further to 0°C and finally freeze, this explains why it takes so long especially for a large lake to freeze even after air temperatures drop well below freezing, and it also explains why the shallower portions tend to freeze first (because a lower total quantity of water needs to be processed by this cycling mechanism).

If ice were more dense than liquid water, then it would form at the surface and sink, and the lake would gradually fill up with ice from the bottom up. There would be no slow conduction step to slow this freezing process down, and it would be much easier for a lake to freeze solid.

4.3.2 Characterization of Flammable Vapor Hazards

1. [Institutional availability of MSDSs.] Identify two chemicals that you believe are in use in some quantity in your institution, at least one of which should not be primarily used in a laboratory setting. Find out if and where your institution has the MSDSs for these chemicals, review the MSDSs, and summarize in one paragraph the nature of the hazards identified there. (Hint: chemicals are used extensively in our homes as well as our institutions, so the identification of such use should not be difficult. MSDSs may be kept in various locations, such as the departments where the materials are used, the organization's safety or security department, the purchasing department, or elsewhere.)

Aside from science laboratory chemicals, here are examples of other classes of materials likely to be encountered in an institutional setting: printing and copying supplies, including chemicals used to clean equipment; physical plant vehicle maintenance supplies (e.g., lubricants, degreasers, fuels); photographic darkroom chemicals; adhesives used in media production and publications; water conditioning chemicals for air conditioning systems or boiler water;

pesticides for landscape or building use; welding supplies and gases; refrigerants used in large refrigeration units; general cleaning chemicals; specialized chemicals used in kitchen operations or cleaning; paints and other coatings.

2. [Library availability of MSDS-type information.] Determine the best sources within your library or, if not readily available there, elsewhere within your institution, of chemical properties information and associated safety information. (Hint: if it would be helpful for you to have specific target chemicals in mind, consider Exercise 6, below, at the same time.) Summarize the availability of sources that provide the values of the parameters discussed in this section for multiple chemicals.

There are many chemical handbooks that summarize data for a wide range of chemicals, some specializing in particular areas such as agricultural or photographic chemicals. A search of a library's holding under the word "chemicals" seems to be one of the most reliable ways to locate these. They are usually in the reference section. A very common one is the Merck Index, a fairly comprehensive collection of data on chemicals and pharmaceuticals. Aside from the library, the institution is likely to have handbooks in the science department, the campus safety department (especially if there is a "life safety" officer), and in any office with overall environmental responsibility. The safety offices in many larger universities also post MSDSs on-line (see Exercise 4).

3. [Vendor availability of MSDSs.] Identify two chemicals that you can buy in your local community and ask the vendors to let you see their MSDSs. Summarize how this process goes and how hard you or they had to look to find these MSDSs.

Based on past student experience, this should not be difficult, although the students may be the first people who ever have asked to see these sheets. Oil dealers, hardware stores, supermarket chains, and many others do indeed have these sheets at their locations.

4. [On-line availability of MSDSs.] By using standard on-line computer search tools, determine the availability of one or more sets of MSDSs that you can access by computer.

Searches under the term "MSDS" or under the name of a particular chemical of interest should turn up numerous sources accessible on-line. General environmental sites also typically have links or citations to on-line sources of MSDSs. Although a number of sites have the identical database, there are often alternative MSDSs or other kinds of toxic data sheets that can help in finding somewhat elusive parameters.

5. [US OSHA regulations regarding MSDSs.] Outline the regulations issued by the Occupational Safety and Health Administration (OSHA) with respect to MSDSs, providing further detail on the subject of who is required to possess MSDSs for chemicals in their possession. (Hint: many libraries have the Code of Federal Regulations, a multivolume series containing the regulations issued by each federal agency. Alternatively, such regulations can be found and searched on-line by computer.)

The regulations are found in the Code of Federal Regulations under the Occupational Safety and Health Administration, available both on-line and in many library reference sections. The particular standard is called the Hazard Communication Standard and applies to the responsibility of employers to make hazard information available to their employees. OSHA has no authority to regulate the availability of such information for the public although there are also regulations of other agencies pertaining to such issues. If employers are going to have access to such information themselves, there needs to be a whole chain of communications right on down from the original manufacturer, through any intermediate distributors, right to the point of purchase. Since "employers" might include people going into hardware stores, say, to buy items to use on a commercial job site, those who sell to them have to be able to provide this information at the time of the sale. This standard defines responsibilities through this chain of communication. In addition to the standard itself, found in 29CFR1910.1200, there is an Appendix E that provides additional advice on how to meet the standards. There is also extensive OSHA correspondence interpreting the issues raised, such as a detailed inquiry by the Ohio Hardware Association and an OSHA response concerning the requirements on retail hardware stores. In general, if someone sells hazardous materials to people who are employers, rather than ordinary consumers, then you have to be able to respond to their requests for material safety data sheets. There are a number of exceptions to the rules, generally in cases where the classes of chemicals are already regulated by other agencies. All this information, including the correspondence and interpretations, is available on-line from OSHA. The first parts of the standard are reproduced below:

1910.1200 - Hazard Communication.
(a) "Purpose." (1) The purpose of this section is to ensure that the hazards of all chemicals produced or imported are evaluated, and that information concerning their hazards is transmitted to employers and employees. This transmittal of information is to be accomplished by means of comprehensive hazard communication programs, which are to include container labeling and other forms of warning, material safety data sheets and employee training.
(2) This occupational safety and health standard is intended to address comprehensively the issue of evaluating the potential hazards of chemicals, and communicating information concerning hazards and appropriate protective measures to employees, and to preempt any legal requirements of a state, or political subdivision of a state, pertaining to this subject. Evaluating the potential hazards of chemicals, and communicating information concerning hazards and appropriate protective measures to employees, may include, for example, but is not limited to, provisions for: developing and maintaining a written hazard communication program for the workplace, including lists of hazardous chemicals present; labeling of containers of chemicals in the workplace, as well as of containers of chemicals being shipped to other workplaces; preparation and distribution of material safety data sheets to employees and downstream employers; and development and implementation of employee training programs regarding hazards of chemicals and protective measures. Under section 18 of the Act, no state or political subdivision of a state may adopt or enforce, through any court or agency, any requirement relating to the issue addressed by this Federal standard, except pursuant to a Federally-approved state plan.
(b) "Scope and application." (1) This section requires chemical manufacturers or importers to assess the hazards of chemicals which they produce or import, and all employers to provide information to their employees about the hazardous chemicals to which they are exposed, by means of a hazard communication program, labels and other forms of warning, material safety data sheets, and information and training. In addition, this section requires distributors to transmit the required information to employers. (Employers who do not produce or import chemicals need only focus on those parts of this rule that deal with establishing a workplace program and communicating information to their workers. Appendix E of this section is a general guide for such employers to help them determine their compliance obligations under the rule.)
(2) This section applies to any chemical which is known to be present in the workplace in such a manner that employees may be exposed under normal conditions of use or in a foreseeable emergency.
(3) This section applies to laboratories only as follows:
(i) Employers shall ensure that labels on incoming containers of hazardous chemicals are not removed or defaced;
(ii) Employers shall maintain any material safety data sheets that are received with incoming shipments of hazardous chemicals, and ensure that they are readily accessible during each workshift to laboratory employees when they are in their work areas;
(iii) Employers shall ensure that laboratory employees are provided information and training in accordance with paragraph (h) of this section, except for the location and availability of the written hazard communication program under paragraph (h)(2)(iii) of this section; and,
(iv) Laboratory employers that ship hazardous chemicals are considered to be either a chemical manufacturer or a distributor under this rule, and thus must ensure that any containers of hazardous chemicals leaving the laboratory are labeled in accordance with paragraph (f)(1) of this section, and that a material safety data sheet is provided to distributors and other employers in accordance with paragraphs (g)(6) and (g)(7) of this section.
(4) In work operations where employees only handle chemicals in sealed containers which are not opened under normal conditions of use (such as are found in marine cargo handling, warehousing, or retail sales), this section applies to these operations only as follows:
..1910.1200(b)(4)(i)
(i) Employers shall ensure that labels on incoming containers of hazardous chemicals are not removed or defaced;
(ii) Employers shall maintain copies of any material safety data sheets that are received with incoming shipments of the sealed containers of hazardous chemicals, shall obtain a material safety data sheet as soon as possible for sealed containers of hazardous chemicals received without a material safety data sheet if an employee requests the material safety data sheet, and shall ensure that the material safety data sheets are readily accessible during each work shift to employees when they are in their work area(s); and,
(iii) Employers shall ensure that employees are provided with information and training in accordance with paragraph (h) of this section (except for the location and availability of the written hazard communication program under paragraph (h)(2)(iii) of this section), to the extent necessary to protect them in the event of a spill or leak of a hazardous chemical from a sealed container.
(5) This section does not require labeling of the following chemicals:
(EXCEPTIONS AND FURTHER DETAILS ARE PRESENTED NEXT)

6. **Using any of the sources of MSDSs or chemical information that you pursued in connection with the above exercises, or any other source of information, fill in as many of the blanks as you can for the chemical properties shown in Table 4-5. For any values that do not seem to be available, indicate why it is reasonable that they are not provided on the MSDSs.**

The values have been provided in Table 4-B. Note that kerosene and turpentine are mixtures of different materials, which is why some of their values show a range. The values reported in the table come from various sources. It is possible that one might find slightly different values using other sources of information, depending on tests or estimates from different manufacturers, whose products may differ slightly. Even the reported molecular weight values, taken here from manufacturers' publications, may differ slightly from the values one would calculate using the atomic weights reported in Table 4-2 in Volume 1. Specific gravity, which of course varies with temperature, is usually reported for typical ambient conditions (68°F or 20°C), even if that assumption is not made explicit, unless the material has a different physical state at that temperature (as with liquid anhydrous ammonia).

> If you happen to be using the CAMEO program for your modeling, the internal database actually gives values for all the indicated table entries. So you may want to introduce the program early, here, and let the students get familiar with it as a data source. If the students are going to seek the information on-line or from other sources, it is almost as easy, because there are alphabetized lists of MSDS sheets on-line and it is easy to access one right after the other. Most have a very similar format and are very easy to read. This problem might be good for a group assignment to let them divide up the effort and keep it from getting too repetitive or tedious. After you have obtained the values in this table, it might be good to talk your way through them in class, commenting on the nature of the flammability risks from the various materials.

7. **Within a given family of chemicals, it is not uncommon to find a general relationship between molecular weight, volatility, and boiling point. For example, the boiling point might a good indicator of volatility because it reflects the fact that more volatile substances, in that they vaporize more easily, are likely to reach a vapor pressure equal to atmospheric at a lower temperature. On the other hand, heavier molecules might correspond to lower volatility because they would require more energy before they can become freed up and leave the liquid state. Explore the validity of this idea by considering the following family of chemicals: methane, ethane, propane, butane, pentane, and hexane. This family, called *alkanes,* all consist of chains of carbon atoms (ranging from 1 to 6 in the above list) surrounded by the maximum complement of hydrogen**

TABLE 4-B
Selected chemicals and their properties to be determined in Exercise 6

Chemical	Molecular Weight	Specific Gravity	Boiling Point		Flash Point		LFL	UFL	Vapor Pressure at about 68°F	Vapor Pressure at alternative temp.
			°F	°C	°F	°C	%	%	(mmHg)	(mmHg)
Acrylonitrile	53.06	.81	172	78	30	−1	3	17	88	
Anhydrous ammonia (liquid)	17.03	.68 (at −28°F)	−28	−33	(A)	(A)	16	25	(B)	400mmHg at 49.72(°F)
Benzene	78.11	.879	176	80	12	−11	1.3	7.9	79.4	
Cyclo-hexane	84.16	.779	177.3	81	−4	−20	1.33	8.35	79.74	
Hexane (normal)	86.17	.659	155.7	69	−7	−22	1.2	7.7	127.58	
Kerosene	170 (approx.)	.8	392 −500	200 −260	100	38	.7	5	2.12	
Nitric Acid (fuming)	63	1.25	181	83	(C)	(C)	(C)	(C)	48	
Turpentine	136 (approx.)	.86	302 −320	150 −160	95	35	.8	(D)	32.89	

(A) Not relevant since it boils well below ambient.

(B) Need to use an alternative temperature below its boiling point for vapor pressure to be meaningful as defined in text.

(C) Not flammable.

(D) Not available, presumably because the volatility is sufficiently low that it is not relevant or likely to be reached.

TABLE 4-C
Variation of properties within the alkane family (Exercise 7)

chemical	molecular weight	boiling point (°F)
methane	16.04	−258.7
ethane	30.07	−127.5
propane	44	−40
butane (n)	58.12	31.1
pentane (n)	72.15	97
hexane (n)	86.17	155.7

atoms that can attach to them under the principles of chemical bonding. (Note: You should be able to find the chemical data you need by the methods practiced above. If you encounter variations on these chemicals in your database, use the values for the "normal" or "(n)" form of the chemical, which has the simplest single-chain molecular structure in each case.)

Lower molecular weight should lead to higher volatility and hence also lower boiling point, since the lighter molecules should be more mobile and hence more able to escape from the liquid. This is supported by the data in Table 4-C. Note that this principle would apply much less reliably as one deals with chemicals that have essentially different kinds of molecular structures. In this latter case, the nature of the attractive forces between molecules can be quite different, thereby aiding or impeding vaporization.

As further background on the chemicals referred to in this problem, Figure 4-A shows a two-dimensional schematic of their molecules. The basic chemical principle controlling the structure of these and other hydrocarbons

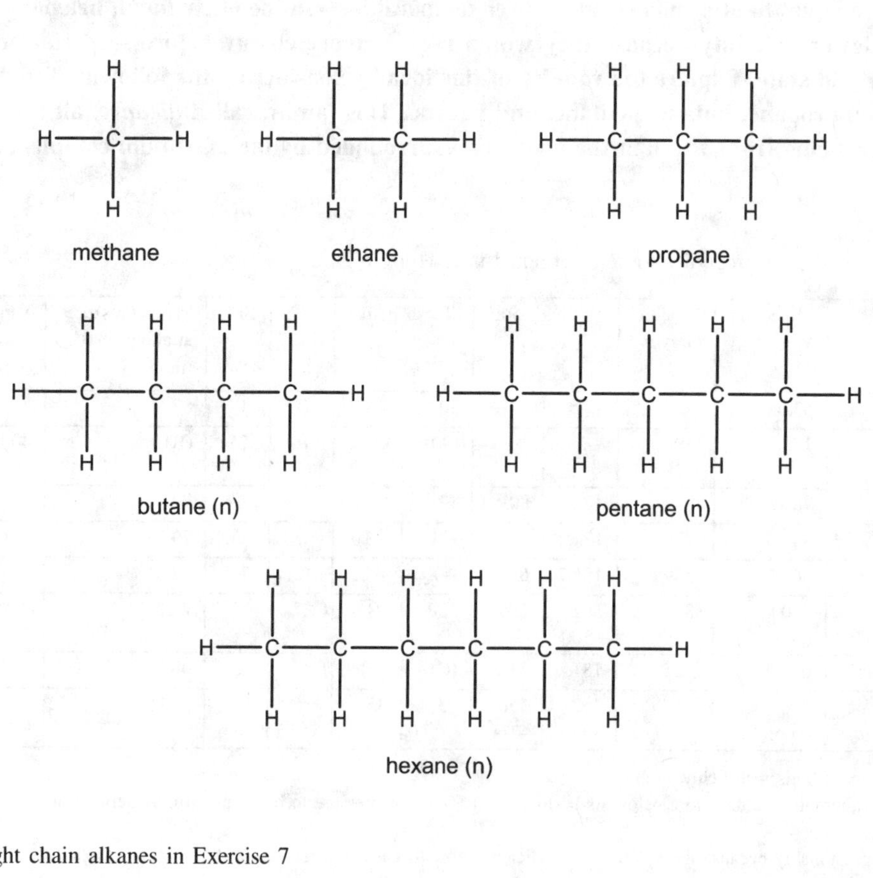

FIGURE 4-A
Structure of straight chain alkanes in Exercise 7

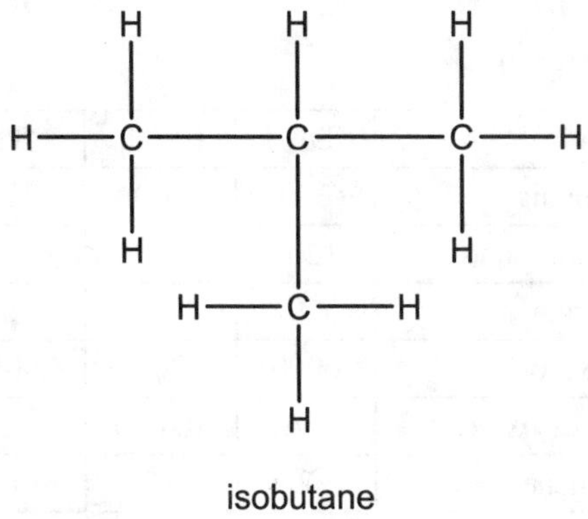

isobutane

FIGURE 4-B
Structure of isobutane (Exercise 7)

is that each carbon atom must have four bonds and each hydrogen atom one bond. Some molecules may have double or triple bonds in some places, which count for that respective number.

This figure shows all straight-chain molecules, but note that in the case of butane and larger members, it is possible to have the same chemical composition but a rearrangement of the chain structure, as shown in Figure 4-B for butane. This leads to a chemical that may have somewhat different chemical properties, so its name must recognize the difference. Thus the straight chain butane is called normal butane, often written "butane(n)," and the variations by various prefixes, such as in isobutane. Looking at Table 4-B for hexane and cyclohexane, this represents still a further variation in that the six-member carbon chain in cyclohexane actually forms a circle closing on itself, in which case the room for two hydrogen atoms is lost, which is why the molecular weight is less in that table. (Try sketching this.)

8.   When the concept of LFL was introduced in the text, it was referred to as a percent "of the molecules." For example, an LFL of 5% would mean that at least 5% of the molecules in the vapor space would need to be those of the flammable material in order to support combustion. However, the MSDSs you have examined probably refer to the LFL as a "percent by volume," a distinction that may or may not have caught your attention. These two concepts are actually essentially the same because of a physical principle applying to "ideal gases" that says that an equal number of gaseous molecules take up the same space under the same conditions, regardless of their mass or chemical composition. Given this principle, consider a vapor space that contains 2% acetone molecules and 3% methane molecules, so that they are in a volume ratio of 2:3. What is the ratio *by mass* or *by weight* of these two components in the vapor space?

The molecular weight of acetone is found to be 58.08, whereas that for methane has already been given in Table 4-C to be 16.04 . Therefore a molecule of acetone is considerably heavier than a molecule of methane. Hence the ratio of their weights could be calculated as:

$$\frac{2 \times 58.08}{3 \times 16.04} = 2.41 = 2.41:1$$

and hence 2.41:1 of acetone to methane.

4.3.3 Characterization of Toxicity Hazards

1.   Search the literature, either on-line or with printed sources, to determine reasonable concentration levels that you might use when trying to assess the risk from airborne concentrations of each of the following: ammonia, carbon monoxide, chlorine, methanol, nitrous oxide, and phosgene. (Hint: don't be surprised if you find a wide variation in the amount of information on MSDSs for the same chemical, but from different

TABLE 4-D
Toxicity measures for selected chemicals (ppm)

Chemical	IDLH	EEGL	TLV
Ammonia	300	100	25–35
Carbon monoxide	1200	400	25
Chlorine	10	3	.5–1
Methanol	6000	200	200–250
Nitrous oxide	N.A.	10,000	50
Phosgene	2	0.2	0.1

manufacturers or different sources. Also, if you are accustomed to on-line information searching, do not completely forget other kinds of sources that may exist at your institution in case you run into difficulty.)

Table 4-D presents several toxicity measures for the chemicals listed in the exercise. The IDLH and TLV values in this case have been taken from the CAMEO database, and the EEGL values are from the ARCHIE manual. Aside from using MSDSs as alternative sources, the IDLH values may also be found in their source publication: the NIOSH *Pocket Guide to Chemical Hazards,* which is periodically updated by the National Institute for Occupational Safety and Health, part of the US Department of Health and Human Services. EEGL values are published by the National Research Council. There are numerous other measures not listed here. However, the point of this entire exercise is not to bring the students to a professional level in knowing the precise advantages and disadvantages of different measures, but rather for them to appreciate that different measures do exist and that there may be disagreement and uncertainty about which to use for a given situation. In addition, they should understand the implications of using alternative values on possibly overestimating or underestimating the risk. (See the next exercise.)

2.   Suppose two different organizations publish emergency action concentration levels for a given chemical. Organization A publishes a value of 10 ppm based on their analysis of the research data, and Organization B publishes a value of 50 ppm based on their interpretation of the same data.

a)   Which organization would you say has the more *conservative* standard? Explain.

b)   If a toxic vapor cloud of this material is forming around a spill and you calculate the extent of the hazard zone of the cloud based on the two different published concentration values, which value will lead you to identify a larger hazard area?

(Note: the identification of hazard levels is somewhat judgmental, generally involving qualitative or semi-quantitative extrapolation from data that may be based on animal studies or very different conditions or concentrations. This is why there can be different interpretations.)

a)   Organization A has the more conservative standard, as they express a level of concern at even lower concentrations than those recommended by Organization B.

b)   The hazard zone based on Organization A's standard will be larger, as it will extend out a farther distance well beyond the concentration of 50 ppm identified by Organization B.

Note: one must obviously be careful not to use an overly conservative value in the hazard calculations, or the results of the analysis will be useless. For example, if you were to use a TLV, which is really a permissible, regular occupational exposure level and then you were to recommend to the fire chief or the mayor that the entire area within this zone be evacuated, you would be encouraging a senseless and perhaps costly and hazardous operation. Therefore the IDLH value would probably be the more reasonable choice. In any case, lacking a specific recommended evacuation threshold from one of the cognizant professional authorities, the IDLH is a reasonable hazard level to focus on. (You will see in the discussion of the modeling that we might then still use half the IDLH to account for variability and uncertainty.)

3. The first two parts of this problem involve specific conversions to and from ppm units. The last part asks you to derive a general formula that could be used for all such problems. You may wish to do the specific parts first to improve your mastery of the logical process, or you may wish to do the third part first and then use that result for the first two parts. (Note: the molecular weight values for these materials should have been determined earlier in Exercise 6 of Section 4.3.2.)

a) Convert a 75-mg/m^3 concentration value for ammonia to ppm units.

b) Convert a 200-ppm benzene concentration to mg/m^3.

c) Consider a chemical whose molecular weight is M. Find a simple equation that can be used to convert between a value A, representing its concentration in air in terms of mg/m^3, and B, the equivalent concentration in ppm units.

Noting all the cancellations in the unit conversion in the text, we obtain the final reciprocal relationships:

$$B = \frac{A}{M} \times 22.4$$
$$A = \frac{MB}{22.4}.$$

Therefore, for part a, the new concentration is

$$B = \frac{75}{17.03} \times 22.4 = 98.6 \text{ ppm.}$$

For part b, we have

$$A = \frac{78.11 \times 200}{22.4} = 697.4 \text{ mg/m}^3.$$

4. In Chapter 3, just at the end of Section 3.7, the ppm unit was briefly introduced in connection with nitrogen oxide concentrations. At that time, you were given simply the conversion relationship that 1 ppm of NO_x was equivalent to 1.1×10^{-7} lb/ft^3. Based on your experience in the current section, can you reconstruct the basis for that conversion factor?

First, we need to convert the quantity 1.1×10^{-7} lb/ft^3 to mg/m^3. This is done with our usual procedures:

$$\frac{1.1 \times 10^{-7} \text{ lb}}{\text{ft}^3} \times \frac{(3.28 \text{ ft})^3}{(1 \text{ m})^3} \times \frac{454.54 \text{ g}}{1 \text{ lb}} \times \frac{1000 \text{ mg}}{1 \text{ g}} = 1.76 \text{ mg/m}^3.$$

Now we use either of the formulas from the last part of the previous exercise to determine what molecular weight M this conversion must correspond to:

$$M = \frac{A}{B} \times 22.4 = \frac{1.76}{1} \times 22.4 = 39.4.$$

This M value is reasonable for nitrogen oxides from the exhaust gases because it is consistent with being an average of the values for the principal components, namely, NO ($M = 34.01$) and NO_2 ($M = 46.01$). (This mix will vary with the combustion source, but the molecular weights for these and other components are sufficiently similar that a value near 40 will generally give very acceptable results.)

5. We have encountered several units for specifying the concentration of a chemical vapor in the air: ppm, lb/ft^3, mg/m^3, and volume percent. Suppose that you have actually measured all of these in the field at a temperature of 68°F and atmospheric pressure of 14.7 psi. Now suppose that night arrives and the temperature of the same air mass drops, so that it becomes more dense, while the atmospheric pressure remains the same. If you were now to repeat your measurements of the concentration expressed in each of the four sets of units listed, which ones would still have the same value, which would change, and in which direction would these latter change?

The ppm and volume percent units would be unchanged because they measure the relative numbers of chemical and air molecules, which factor is not affected if a sample is just "squeezed" into a smaller volume. The other two values would increase, because for a given sample of the air mass, the numerator (amount of chemical) would stay the same and the denominator (volume) would decrease.

Incidentally, while this problem can be answered on the basis of ordinary intuition, for science-oriented students you may want to mention the ideal gas law, $PV = nRT$, which relates the pressure, volume, absolute temperature, and number (n) of moles present, by means of the universal gas constant R. (See a chemistry book for more details.) In this case, P and n are unchanged, T decreases, and hence V must decrease. Since many people will have encountered this in basic physics or chemistry courses, it is nice to be able to introduce it in this very concrete setting. In fact, it even applies to each component of a mixture of ("ideal") gases, where P would then be the corresponding partial pressure, V still the total volume, and n the number of moles of the component under consideration.

4.4 Typical Quantitative Issues

1. **For either your hometown (if in the US) or for the town or city in which your institution is located, find out if there is a "local emergency planning committee" (LEPC) or equivalent organization responsible for preparedness for hazardous materials emergencies. Summarize their scope of activities. Find out to what extent they use mathematical or computer modeling in their work.**

Many communities have such committees, although in some cases the responsibilities have been kept primarily at the state level. Typical activities include: meetings (perhaps quarterly or less), occasional practice exercises (in the field or "table-top"), application for grants from state or Federal levels to support some activities (e.g., overtime pay for training programs, upgraded equipment), discussion of priorities for attention in the community (e.g., troublesome plants, transportation through the community), and relationships among agencies and organizations who would need to work in a coordinated way on a major hazmat incident. Computer modeling tools have been made available to most of these committees and training provided either with EPA support or from private industry. Some have active modeling programs intended to understand better local sources of risk, and others have no such programs. The program in the author's school's community, Waltham, MA, is very active. This is a small city in the Boston suburbs with a long history of industrial activity. This interest preceded the role that the college has taken to support the program with student internships, computer support, and faculty involvement.

> Note in connection with this previous response the potential involvement of students and faculty in LEPC activities, as well as in corresponding activities at the state or Federal level. For example, students can perform model calculations on community-relevant sample scenarios designed by emergency personnel and faculty. Such scenarios can and have been used as the basis for actual practice exercises (see example later in the text). For the next exercise, you could have students contact different departments, or you could arrange a tour or class visit in conjunction with the fire department.

2. **Contact a local fire department to inquire about the degree to which they have equipment and procedures for dealing with hazardous materials emergencies. Summarize their observations on the potential for such incidents in the local community and what they consider to be the largest sources of risk. (Note: an individual or class visit to a fire department, especially a "hazmat" unit, can be very interesting and would touch on many of the topics to be discussed in this chapter. Fire departments usually welcome such interactions with members of the community.)**

Specialized equipment, except for large cities, is usually maintained on a regional basis. This may include a special hazmat unit truck containing the full range of items that may be needed in a response. Examples include: special clothing (some resembling "moon suits") at different levels of protection, field sampling kits for air concentrations of many substances and of combustible gases, air supplies, equipment for sealing leaks (patches, etc.), mechanical tools, communications and lighting equipment, extensive databooks with chemical properties and advice on remedial actions, computers (for data retrieval and/or modeling), health monitoring equipment, decontamination equipment (for washing down firefighters or others who have been exposed), and miscellaneous fire-fighting gear. During tours of hazmat facilities or on-campus demonstrations, students generally enjoy trying on the special gear, an exercise that makes the subject more real to them and that also lets them appreciate how difficult it is to work when encumbered by such equipment.

4.5 Structure and Use of Hazmat Computer Modeling Packages

(No exercises.)

4.6 The Analysis of Typical Scenarios

Which computer program to use?

Here are some further comments on the choice of a modeling package. These comments apply as of the date of this writing (1998). ARCHIE and CAMEO are the two programs that are most directed to a more general audience, being developed for use of LEPC's or other local personnel who may be involved in emergency planning or response. They compare in many respects like a traditional Volkswagen and a Cadillac compare.

ARCHIE comes on one disk; you run it from DOS, and you don't really need a manual. There is likely to be a little frustration at first because it is hard to turn around within the program and change values if you make a mistake; you generally have to just "feed it junk" at the rest of the prompts until it finally offers you the option to start over or make changes. It also has the annoying habit of asking you for confirmation of each input, but you soon catch on that the "Enter" key is the expected default response, so you don't have to keep using the Y or N keys to answer each time. There is no internal database, so you have to get your chemical data elsewhere, and there is no mapping program, so you get only tabular output. The program is public domain and free. There is a user's manual, which is an excellent background document on hazmat issues, but you don't need it. (It might be good for the teacher to have a copy for further background.) Some of the specific computer-related aspects (like the logical interaction of submodels) can be accessed from a program menu. The program is used by practitioners, and there are on-line discussion groups for such users. It is not being promoted aggressively by its sponsors and is likely to remain a DOS program.

On the other hand, CAMEO comes on six high-density diskettes, with additional diskettes for the associated ALOHA and MARPLOT programs. You would need ALOHA for vapor dispersion calculations, the main kind of scenarios we will be analyzing. MARPLOT permits automatic plotting of hazard zones on computerized maps. CAMEO has an extensive internal database in a consistent format and with the parameters needed for the modeling. Documentation is extensive, and it is much more difficult to go directly to the program and use it without reviewing some of this documentation. CAMEO is set up to handle some specific data management requirements of companies with respect to toxic chemical inventories, one reason that the program is so much larger than ARCHIE. However, these additional capabilities are not necessary for our work. CAMEO is being actively promoted by EPA and other organizations and is likely to continue to be upgraded. There are training sessions offered to LEPCs and others. Although developed for government agencies, the program is not available free. However, there are academic licensing fees that are much less than commercial rates, and there may be other ways to obtain experience or access, such as through involvement with an LEPC or for certain research purposes. This is a Windows program, and there is also an earlier version for Macintosh systems.

A possible advantage of using CAMEO is that it may open up some internship or employment opportunities for students, although we have also had students experienced with ARCHIE work as interns to help a fire department implement CAMEO.

The solutions reported below are based on ARCHIE, since it is expected that logistical and cost considerations will make that the more popular choice of software.

There are also excellent commercial packages for performing these kinds of calculations, complete with internal databases, and it may be that a local consulting company or chemical company that produces or uses such models may be able to make such a package available for class use.

Model peculiarities

You should also be warned that all models, under certain combinations of input values, are likely to give slightly peculiar results. For example, a model might decide internally whether a release should be modeled as a plume or a puff. Say the criterion is whether the duration of release is greater than or less than 5 minutes. It might happen that a release of duration 4.9 minutes actually shows a much smaller value than a release of 5.1 minutes, or even vice versa, because the simplified submodels used to define the cases don't match up perfectly at the boundary. This happens to some extent with almost every model, and it is more likely to happen with ones that have many simplifications built in so as to make it easier for people to use them. You may come across these inconsistencies in some of your calculations, but keep in mind that the models used here are not intended to produce precise results, but rather more general indications of hazard distances and other related factors. We will present results in seemingly precise terms just to make it easier to make comparisons, but their real-life application would be in more general terms.

Caution about not assigning too many of these exercises at once

This exercise set really represents the culmination of the chapter; and over the long run, the students should probably do all of the exercises. However, it might be wise to assign them in different ways and only a few at a time. For example, you might want to have them work on one problem in groups while still in the classroom (assuming they have computer access in the classroom, such as if one notebook computer per group can be obtained), so that you can go around and troubleshoot as they initially muddle through menus (like all of us!). Then a couple of homework problems might be good to make sure they are on the right track and that they are finding access to the needed input data if it has not been determined in earlier problems. The solutions then could be gone over in class to hammer out any difficulties before going on to further problems. This leisurely and thoughtful approach is more likely to leave them with insights that they will carry away, and it is more likely to provoke questions on their part about this aspect of environmental protection.

These exercises require you to use a modeling package for hazmat emergency situations, as well as to have access to data for the chemicals under discussion. Some models contain internal databases that will meet your needs, whereas others require you to obtain the desired information from reference sources such as MSDSs. Such sources have been discussed further in previous sections. See Section 4.7 for further discussion of available modeling packages. Be very careful with your units and unit conversions in applying data to the models. In addition, be sure to check the reasonableness of your assumptions and model results at every step of the calculations. In some problems, you will find that you need to make additional assumptions beyond the information stated in the problem. State your assumptions clearly in these cases, along with a brief rationale about why you believe them to be realistic.

1. **Estimate how long it would take to empty a vertical cylindrical tank 40 feet in diameter and 50 feet high. The tank contains #2 diesel fuel and is half full at the outset. It is being discharged from a broken 4-inch pipe at the bottom that was accidentally hit by a payloader working in the area. You can assume the outside temperature is around 68°F.**

Model calculations indicate that it would take about 482.6 minutes to discharge. The precise output value will be given in these and other exercises to enable comparisons with calculations by others, but naturally the value should be interpreted in an approximate sense. Thus the tank would take about 8 hours to empty if the leak continued unabated.

The ARCHIE printout of results is reproduced below to aid comparison with other calculations, and it includes at the end a summary of the input values and assumptions. This is an example of a situation, as discussed in the text, where you have to provide this particular model with additional input values that are not relevant to the particular output you are seeking. (The model anticipates that you will be going further into the risk situation for the scenario.) In these situations, I generally give spurious values and indicate these by using numbers with repeated digits within the range acceptable to the model. This makes it easier for me to identify these values in case I return later to carry the scenario further, when I may want to input accurate values that will actually be used. You can see these in the input table below for the following parameters: normal boiling point, molecular weight, temperature of container contents, and ambient temperature. Changes in these input values have no effect on the time to empty the tank. (If viscosity effects were included in the model, then the temperature value for the tank contents would indeed be relevant. However, viscosity effects are not included in this model.)

ARCHIE printout for this problem:

```
HAZARDOUS MATERIAL    = diesel
DATE OF ASSESSMENT    =
NAME OF DISK FILE     = EX1S4-6.ASF

*** SCENARIO DESCRIPTION

    Empty time for tank only

******* DISCHARGE RATE/DURATION ESTIMATES

        Liquid discharge from nonpressurized container

        Average discharge rate = 3452.9      lbs/min
        Duration of discharge  = 482.6        minutes
        Amount discharged      = 1666300      lbs
        State of material      = Liquid


                INPUT PARAMETER SUMMARY

        - - - - - - - - - - - - - - - - - - - - - - - - - - - - - - -

PHYSIOCHEMICAL PROPERTIES OF MATERIAL
    NORMAL BOILING POINT        = 333          degrees F
    MOLECULAR WEIGHT            = 222
    LIQUID SPECIFIC GRAVITY     = .85

CONTAINER CHARACTERISTICS
    CONTAINER TYPE              = Vertical cylinder
    TANK DIAMETER               = 40           feet
    TOTAL WEIGHT OF CONTENTS    = 1666300      lbs
    WEIGHT OF LIQUID            = 1666300      lbs
    LIQUID HEIGHT IN CONTAINER  = 25           feet
    WEIGHT OF GAS UNDER PRESSURE = 0           lbs
    TOTAL CONTAINER VOLUME      = 62831        ft3
                                = 470038       gals
    LIQUID VOLUME IN CONTAINER  = 31416        ft3
                                = 235023       gals
    DISCHARGE HOLE DIAMETER     = 4            inch(es)
    DISCHARGE COEFFICIENT OF HOLE = .62
    TEMP OF CONTAINER CONTENTS  = 222          degrees F

ENVIRONMENTAL/LOCATION CHARACTERISTICS
    AMBIENT TEMPERATURE         = 99           degrees F

KEY RESULTS PROVIDED BY USER INSTEAD OF BY EVALUATION METHODS
    NONE OBSERVED

KEY RESULTS OVERRIDDEN BY USER AT SOME POINT AFTER COMPUTATION
    NONE OBSERVED
```

2.   A westbound gasoline tank truck overturns on Interstate I-44, and its cargo of 7,000 gallons spills out almost immediately and collects in a pool in a depression just off the right shoulder of the road. This pool of gasoline is roughly circular and has a diameter of 20 feet, and it is located off the side of the westbound lanes, about 25 feet at its nearest point from the westbound roadway and 150 feet from the eastbound lanes. There is a slight wind blowing from the north at 4 mph, and it is very sunny. You are the local fire chief, and you arrive on the scene within 10 minutes of the accident. Your first decision has to do with whether to close down the *eastbound* side of the road for fear of ignition of the vapors from the evaporating pool by a passing car. It's a hot day (80 degrees F), and the smell of gasoline is very strong. Is this a significant hazard? (Hint: gasoline is a mixture of various chemicals, each with its own physical and chemical parameters. Table 4-10 provides certain composite modeling parameters for this common material that have been found by practitioners to give good equivalent hazard analysis results.)

The calculated downwind distance to a concentration of 1/2 LFL is 66 feet. The model treats the actual release as a point release, using the area of the pool to determine the evaporation rate. However, this 66-foot distance should be treated as from the downwind edge of the pool, as any other approach is non-conservative. This is considerably less than the distance to the eastbound lanes, so closing of those lanes would not be strictly necessary on the basis of this information. I would take this as the intended and a perfectly acceptable answer.

However, depending on the sophistication of your class, you may want to discuss or have them investigate this answer further along the following lines. In most modeling situations, one would also want to think about whether changes in conditions could significantly change one's conclusion, which is one of the reasons why several of the later exercises (as well as some of the exercises on air pollution modeling in Chapter 3), help one to develop some general intuitive principles about the sensitivity of model output to the input assumptions. For example, if the weather clouded up or the wind speed changed, could the risk change appreciably? First, keep in mind that the time frame for response to an incident like this is probably not long, and because the scenario is relatively simple and the material is well known to the fire department, they would quite quickly probably spray foam on the surface to reduce the rate of evaporation, while the next stages of cleanup were being planned (probably pumping from the pool into a clean-up contractor's tank truck). So even if some clouds obstructed the sun, the ground would still be warm and the instability effects of strong sun would still be operative. But the wind could change or fluctuate, so a reasonable sensitivity analysis to this would include increasing the windspeed over a reasonable range, still only using the stability class from the first column (bright sun) part of the stability class table. Since the model actually uses a "heavy gas" submodel for gasoline vapors, which are heavier than air, some of the results are slightly different than what you would expect from Gaussian dispersion, as in Chapter 3. In any case, the hazard distances do indeed increase, although not to the 150-foot level. For example, with an 11.2-mph wind and the corresponding C stability level, the hazard distance to 1/2 LFL is 100 feet.

The ARCHIE output for the baseline scenario is reproduced below. Note the use of some spurious input values (such as tank dimensions) to satisfy the program's needs, even though the use of a one-minute release scenario renders them irrelevant to the final answer. This issue was discussed in the text.

```
HAZARDOUS MATERIAL     = gasoline
DATE OF ASSESSMENT     = 5/1/97
NAME OF DISK FILE      = EX2S4-6.ASF

*** SCENARIO DESCRIPTION
     truck overturns on interstate

****** DISCHARGE RATE/DURATION ESTIMATES
          All contents assumed by user to discharge within 1.0 minute
          Duration of discharge   = 1            minutes
          Amount discharged       = 37368        lbs
          State of material       = Liquid
```

******* LIQUID POOL SIZE ESTIMATES

    Evaporating pool area   = 314   ft2
    Burning pool area   = 314   ft2

  Note: Pool is assumed to ignite after pool achieves maximum size.

******* LIQUID POOL EVAPORATION RATE/DURATION ESTIMATES

    Vapor evolution rate   = 30.3   lbs/min
    Evolution duration   = 1234.3   minutes

******* POOL FIRE HAZARD ESTIMATION RESULTS

    Burning pool radius   = 10   feet
    Flame height   = 49   feet
    Fatality zone radius   = 37   feet
    Injury zone radius   = 53   feet

******* FLAMMABLE VAPOR CLOUD HAZARD RESULTS

    For concentration of   1/2 LFL   LFL
                                 ------   -----

    Downwind hazard distance   = 66   46   feet
    Max hazard zone width   = 33   23   feet
    Max weight explosive gas   = 5.7   3.9   lbs
    Relative gas/air density   = 2.2   2.2   initially
    Model used in analysis   = Heavy gas

  Note: Clouds or plumes containing less than 1000 pounds of vapor or gas are
        very unlikely to explode when completely unconfined, except when one
        of a certain few materials have been discharged.

                        INPUT PARAMETER SUMMARY
    -

PHYSIOCHEMICAL PROPERTIES OF MATERIAL
    NORMAL BOILING POINT   = 114.9   degrees F
    MOLECULAR WEIGHT   = 90.9
    LIQUID SPECIFIC GRAVITY   = .64
    VAPOR PRES AT CONTAINER TEMP = 8.186   psia
                                  = 423.6   mm Hg
    VAPOR PRES AT AMBIENT TEMP   = 8.19   psia
                                  = 423.6   mm Hg
    LOWER FLAMMABLE LIMIT (LFL)   = 1.4   vol%
CONTAINER CHARACTERISTICS
    CONTAINER TYPE   = Horizontal cylinder
    TANK DIAMETER   = 11   feet
    TANK LENGTH   = 99   feet
    TOTAL WEIGHT OF CONTENTS   = 37368   lbs
    WEIGHT OF LIQUID   = 37368   lbs
    LIQUID HEIGHT IN CONTAINER   = 1.8   feet
    WEIGHT OF GAS UNDER PRESSURE = 0   lbs
    TOTAL CONTAINER VOLUME   = 9408   ft3
                                  = 70381   gals
    LIQUID VOLUME IN CONTAINER   = 935.8   ft3
                                  = 7000   gals
    TEMP OF CONTAINER CONTENTS   = 80   degrees F

```
ENVIRONMENTAL/LOCATION CHARACTERISTICS
   AMBIENT TEMPERATURE        = 80              degrees F
   WIND VELOCITY              = 4               mph
   LIQUID CONFINEMENT AREA    = 314             sq ft
KEY RESULTS PROVIDED BY USER INSTEAD OF BY EVALUATION METHODS
   NONE OBSERVED

KEY RESULTS OVERRIDDEN BY USER AT SOME POINT AFTER COMPUTATION
   NONE OBSERVED
```

3.   You have recently moved to Connecticut to take over the management of a municipal sports center which includes an ice rink. Your staff tells you that there has long been a problem of refrigeration breakdowns, so you decide to request funds from the town to upgrade the aging refrigeration system, which uses ammonia as its refrigerant. During the public budget review process, several residents who had been unaware that the plant had a large ammonia inventory express concern about this system, not only because of their own proximity but because there is a new nursing home and rehabilitation center located just about a mile away. (At this point, you may be beginning to wish that you had never even suggested any changes to the system.)

Therefore the town decides to do a more thorough review of options for the plant, including closing the rink and contracting ice time from a newer rink in the next town, upgrading the ammonia system according to your recommendations, or replacing the system with a freon-based system. A consultant is hired to help evaluate the risks of the second option, and in a scoping meeting with the fire chief it is decided to define a "worst case" spill scenario as the short time frame (say, one minute) release of the full contents of one of three liquid ammonia receiving tanks on the high pressure part of the cooling circuit, and the further assumption that the material boils off instantly as it is released. Each such tank holds 80 gallons. Furthermore, even though this part of the system is located in a closed room on the back of the building, it has a large garage door that would probably be open during the kinds of maintenance and repair activities that might lead to such an accidental release, so the release should be assumed to be directly to the outside air.

Could the toxic impact of such a release impact a location as far away as the nursing home, in which case you would probably also be responsible for developing emergency procedures for that facility?

Note: this is a typical framework for such a question, but you will find that not all the input data for the model have been provided. In this situation, you must make reasonable assumptions, generally trying to err on the conservative side if necessary. In addition, you need to be sure to find a way to have the model simulate this one-minute release and vaporization scenario.

Since ammonia is a "low boiler," it will boil rapidly upon release, and thus the assumption that the conversion to vapor is essentially instantaneous is quite reasonable, at least for this relatively small quantity. (If it were a larger quantity, then the availability of heat to keep it boiling might be an issue. This is discussed further in Chapter 7.) For this problem you are not asked to investigate flammability aspects, for toxicity risk from ammonia is much greater. In fact, it is quite difficult to get ammonia to ignite or keep burning under the conditions of this problem. The easiest way to incorporate this assumption, at least within ARCHIE, is to go straight to the toxic vapor submodel, skipping the previous ones, and input a hand-calculated vapor evolution rate of 454 pounds per minute, obtained by converting 80 gal/min using the density of liquid ammonia. (Its specific gravity is .68, so its weight per cubic foot is 68% that of water, or $.68 \times 62.4 = 42.43$.) This issue of forcing models to fit the situation was discussed in the text, and it is good for the students to get some practice with it because it really makes them think. In reviewing incorrect answers to this problem, you will probably find releases of 80 gallons of gas, which has a minuscule mass, and similar errors that lead to grossly underestimated hazard zones.

Students probably learn more by getting these things wrong at first than if they stumble onto the right approach immediately at the outset. (But this is why you have to give them only a few problems at a time.)

The IDLH value for ammonia is 300 ppm (determined earlier in the solution to Exercise 1 of Section 4.3.3), so we will use a threshold value of half this amount, or 150 ppm, in our calculations.

TABLE 4-E
Results of ammonia spill sensitivity runs

Stability class	Wind speed (mph)	Hazard distance (feet)
C	8	2025
	12	1726
D	7	3747
	12	3133
	15	2764
E	7	5615
	11	4932
F	5	9889

The critical additional information that you need to provide for this problem involves the weather assumptions. The question does not ask for absolute "worst case" conditions, but only whether there could be risk to the residents one mile away under presumably credible conditions. Naturally, to be conservative, we should be sure to look for unfavorable conditions to input to the model. Since our experience in Chapter 3 showed a balancing effect between higher short distance risk in very unstable conditions and lesser but more distant risks at stable conditions, we will begin our calculations by assuming neutral or D stability. Furthermore, we also know from Chapter 3 that at least for Gaussian plume models and elevated sources, lower winds for a given stability class usually produce higher concentrations since there is less air passing by in the wind for dilution. (In this case, the release would be closer to a puff than a plume.) So now we approach the numerical experiment with D stability and, say, a typical wind for this class of, say, 12 mph.

Other required input parameters have been determined in an earlier exercise and are shown in Table 4-B.

The model calculates a toxic hazard distance directly downwind of 3,133 feet for the particular weather conditions described above. (Naturally, students or others might start with different values.) Now we proceed to repeat the modeling, testing out other assumptions about the weather, to see if the nursing home might fall within the hazard zone.

Table 4-E presents the results from a number of such runs, in which you can easily see some not unexpected trends, and in which you can also see that under certain conditions, particularly at night, the nursing home could indeed lie within the toxic hazard zone.

Note that the duration of the hazard would be expected to be small, due to the instantaneous release. Emergency planning procedures, if required, might involve an automatic alarm initiated at the ammonia storage location and training of nursing home staff to shut windows and turn off ventilation systems at the sound of such an alarm.

The ARCHIE output for the baseline scenario follows.

```
HAZARDOUS MATERIAL    = ammonia
NAME OF DISK FILE     = EX3S4-6.ASF


***    SCENARIO DESCRIPTION
       receiver instantaneous release


******    TOXIC VAPOR DISPERSION ANALYSIS RESULTS
          Downwind distance to concentration of 150 ppm
          -- at groundlevel  = 3133   feet

  Note:   Minimum computable answer is 33 feet! Actual hazard distance may be less.
          See attached table(s) for further details.
```

TOXIC VAPOR DISPERSION ANALYSIS RESULTS

Downwind Distance		Groundlevel Concentration	Source Height Concentration	Initial Evacuation Zone Width*
(feet)	(miles)	(ppm)	(ppm)	(feet)
100	.02	75413	75413	73
317	.06	8502	8502	240
534	.11	3282	3282	390
750	.15	1782	1782	550
967	.19	1138	1138	710
1183	.23	799	799	870
1400	.27	597	597	1020
1617	.31	466	466	1180
1833	.35	376	376	1340
2050	.39	311	311	1500
2266	.43	262	262	1650
2483	.48	224	224	1810
2700	.52	195	195	1970
2916	.56	170	170	2130
3133	.6	150	150	1

*Usually safe for < 1 hour release. Longer releases or sudden wind shifts may require a larger width or different direction for the evacuation zone. See Chapters 3 and 12 of the guide for details. Source height specified by the user for this scenario was 0 feet.

TOXIC VAPOR DISPERSION ANALYSIS RESULTS

Downwind Distance		Contaminant Arrival Time at Downwind Location	Contaminant Departure Time at Downwind Location
(feet)	(miles)	(minutes)	(minutes)
100	.02	.1	1.2
317	.06	.4	1.7
534	.11	.6	2.1
750	.15	.8	2.5
967	.19	1	2.9
1183	.23	1.2	3.3
1400	.27	1.4	3.7
1617	.31	1.6	4.1
1833	.35	1.8	4.5
2050	.39	2	4.9
2266	.43	2.2	5.3
2483	.48	2.4	5.8
2700	.52	2.6	6.2
2916	.56	2.8	6.6
3133	.6	3	7

CAUTION: See guide for assumptions used in estimating these times.

INPUT PARAMETER SUMMARY

- -

PHYSIOCHEMICAL PROPERTIES OF MATERIAL
    MOLECULAR WEIGHT   = 17.03
    TOXIC VAPOR LIMIT   = 150   ppm

ENVIRONMENTAL/LOCATION CHARACTERISTICS
    AMBIENT TEMPERATURE   = 80   degrees F
    WIND VELOCITY   = 12   mph

```
ATMOSPHERIC STABILITY CLASS   = D
VAPOR/GAS DISCHARGE HEIGHT    = 0          feet
```

KEY RESULTS PROVIDED BY USER INSTEAD OF BY EVALUATION METHODS
```
  VAPOR EVOLUTION RATE          = 453.79   lb/min
  VAPOR EVOLUTION DURATION      = 1        minutes
```

KEY RESULTS OVERRIDDEN BY USER AT SOME POINT AFTER COMPUTATION
  NONE OBSERVED

4. A 5,000-gallon tank truck of concentrated nitric acid is pulled over into a roadside rest area by a state trooper who is responding to a cellular phone call from a concerned motorist who noticed that it was dripping liquid along the highway. Upon inspection, it appears that a faulty valve or cracked fitting is the problem, and the material leak rate seems to be increasing and looks like it's now up to about 5 gallons per minute. What is the nature of the hazard and the size of the hazard zone? Assume moderate sunshine with scattered clouds, a temperature of 75°F, and a slight breeze of 4 mph.

As you get into this scenario, you will need to specify the duration of release, which is a missing but important parameter. Of course the absolute maximum would be the total capacity of the tank truck divided by the leak rate, but this would come to 1000 minutes, or almost 17 hours. This is not reasonable, as surely emergency crews could stop the leak well before that. Therefore our baseline run is based on a duration of 3 hours and the maximum credible pool that could form during that period. The resulting hazard distance is 1,255 feet.

If you decided to use a shorter leak duration, assuming emergency action would be taken faster (not unreasonable, but not guaranteed), then you would find (at least using ARCHIE) that the hazard distance would still be about the same.

The ARCHIE results are given below. The IDLH value for nitric acid is 25ppm, and so a threshold of one-half of this value, or 12.5ppm, was used in these calculations.

```
HAZARDOUS MATERIAL    = nitric acid (conc.)
NAME OF DISK FILE     = EX4S4-6.ASF

 ***   SCENARIO DESCRIPTION
       steady leak from tank truck in rest area

 *******   LIQUID POOL SIZE ESTIMATES

       Evaporating pool area   = 8360   ft2

 *******   LIQUID POOL EVAPORATION RATE/DURATION ESTIMATES

       Vapor evolution rate   = 62.6   lbs/min
       Evolution duration     = 180    minutes

 *******   TOXIC VAPOR DISPERSION ANALYSIS RESULTS

       Downwind distance to concentration of 12.5 ppm
       -- at groundlevel      = 1255   feet

  Note:  Minimum computable answer is 33 feet!
         Actual hazard distance may be less.

         See attached table(s) for further details.
         (TABLE OMITTED.)

         INPUT PARAMETER SUMMARY
- - - - - - - - - - - - - - - - - - - - - - - - - - - - - - - - - - -

PHYSIOCHEMICAL PROPERTIES OF MATERIAL
  NORMAL BOILING POINT   = 248.9   degrees F
  MOLECULAR WEIGHT       = 63
```

```
LIQUID SPECIFIC GRAVITY        = 1.5
VAPOR PRES AT CONTAINER TEMP   = 1.065   psia
                               = 55.1    mm Hg
VAPOR PRES AT AMBIENT TEMP     = 1.07    psia
                               = 55.1    mm Hg
TOXIC VAPOR LIMIT              = 12.5    ppm
CONTAINER CHARACTERISTICS
  TEMP OF CONTAINER CONTENTS   = 75      degrees F
ENVIRONMENTAL/LOCATION CHARACTERISTICS
  AMBIENT TEMPERATURE          = 75      degrees F
  WIND VELOCITY                = 4       mph
  ATMOSPHERIC STABILITY CLASS  = B
  LIQUID CONFINEMENT AREA      = NONE
  VAPOR/GAS DISCHARGE HEIGHT   = 0       feet
KEY RESULTS PROVIDED BY USER INSTEAD OF BY EVALUATION METHODS
  DISCHARGE RATE               = 62.55   lb/min
  DURATION DISCHARGE           = 180     minutes
  AMOUNT DISCHARGED            = 11259   lbs
KEY RESULTS OVERRIDDEN BY USER AT SOME POINT AFTER COMPUTATION
  NONE OBSERVED
```

> The next exercise asks the students to use their understanding of basic principles and relationships to estimate the impact on the answers to Example 1 of certain changes in the original assumptions. The problem that follows it then asks them to redo the actual model calculations for each such case and determine the quantitative impact predicted by the model. One way to use this pair of exercises is to do the first as a group exercise in class, comparing and debating proposed answers after each group has a chance to make its decisions. The second problem could then be given as a homework problem. Problems 7 and 8 form a similar pair, involving the more complex acrylonitrile scenario from Example 2 in the text.

5.   With respect to Example 1 in the text (not Exercise 1), provide your best guess about the answers to the following questions, and provide a supporting argument for each. You are not being asked to rerun the model for these situations. In particular, what would you roughly estimate to be the type of effect on the hazard zones for toxic and flammable vapors of each of the following individual changes from the conditions of the original example?

a)   Increasing the broken hose diameter to 4 inches.

No effect, since the pool area is still constrained by the dike. (The pool would of course get deeper faster, but this does not affect the hazard distance.)

b)   Having the tank 95% full at the outset of the leak.

No effect, since the pool area is still constrained by the dike. (The pool could of course get deeper.)

c)   Reducing the temperature to 50°F.

Evaporation should be slower, and therefore less vapor should be diluted by the same wind, hence concentrations should be less. This should reduce the hazard distance.

d)   Eliminating the dike and assuming the ground is level.

This would increase the pool size, leading to a faster evaporation, hence more concentration in the air, and hence a larger hazard distance.

e)   Basing the hazard zones on 10% of the LFL and toxic thresholds.

This should increase the hazard zone because now even a lower, more distant concentration would be considered hazardous.

f) Changing to overcast conditions.

This should increase the stability class of the system, holding the plume together longer. Since the release is at ground level, there is no concentration peak downwind; the point of maximum concentration is at the source (unlike the air pollution situation with a raised stack release.) The hazard distances should increase.

g) Decreasing the windspeed to 2 mph.

This should certainly increase the concentration as long as the stability class remains the same (since at least for Gaussian dispersion, the stability class determines the dispersion factor as a function of downwind distance, and the windspeed only affects the dilution rate at the point of release). Since Example A was carried out using stability class A, this is indeed the situation. Hence the hazard distance should increase.

h) Increasing the windspeed to 10 mph.

Such a change increases dilution at the source but also increases the evaporation rate and changes the stability class to B, which reduces dispersion and holds the plume together for greater distances. It is quite difficult to predict how these effects will balance out.

6. For each of the situations in the previous exercise, carry through the required model calculations and compare your answers to your prediction. Provide a physically plausible explanation for any differences from your expectations as stated in the previous exercise.

a) Increasing the broken hose diameter to 4 inches.

Same calculated hazard distances.

b) Having the tank 95% full at the outset of the leak.

Same calculated hazard distances.

c) Reducing the temperature to 50°F.

The flammable hazard distance decreases from 79′ to 75′. The toxic hazard distance decreases from 102′ to 98′.

d) Eliminating the dike and assuming the ground is level.

The flammable hazard distance increases from 79′ to 141′. The toxic hazard distance increases from 102′ to 184′.

e) Basing the hazard zones on 10% of the LFL and toxic thresholds.

The flammable hazard distance increases from 79′ to 180′. The toxic hazard distance increases from 102′ to 235′.

f) Changing to overcast conditions.

This changes the stability class to D. The flammable hazard distance increases from 79′ to 258′. The toxic hazard distance increases from 102′ to 339′.

g) Decreasing the windspeed to 2 mph.

The flammable hazard distance increases from 79′ to 88′. The toxic hazard distance increases from 102′ to 115′.

h) Increasing the windspeed to 10 mph.

This changes the stability class to B. The flammable hazard distance increases from 79′ to 103′. The toxic hazard distance increases from 102′ to 133′.

7. With respect to Example 2 in the text, provide your best guess about the answers to the following questions, and provide a supporting argument for each. You are not being asked to rerun the model for these situations. In particular, what would be the nature of the estimated effect on the toxic hazard zones for hydrogen cyanide only (a combustion product of acrylonitrile) of each of the following individual changes from the conditions of the original example?

a)   Increasing the yield factor of HCN to 50% from the combustion process.

This should inject five times as much HCN per minute into the air as in the baseline case and hence should increase the hazard distance.

b)   Assuming that ignition takes place only after the pool reaches maximum size.

The baseline example assumed burning at the time and rate of release from the container. If there is no ignition until all the material has leaked out and a huge pool has formed on the surface of the water, then there will be a much more rapid burn rate and higher concentration released into the air in a short amount of time. However, the effective height of release will be higher since the flame will be bigger. Thus the longer diffusion distance back to ground level will somewhat compensate for the higher source term for the diffusion process. However, due to the massive increase in source term, it is likely that the hazard distance will increase.

c)   Changing the weather conditions to moderate sunshine.

This will change the stability from D to B, leading to more diffusion at shorter distances. While this may increase some short distance concentrations, it is likely to decrease the hazard distance, similar to the effects seen in Exercise 6f (although there we were going from a lower stability class to a higher one).

8. For each of the situations in the previous exercise, carry through the required model calculations, and compare your answers to your prediction. Provide a physically plausible explanation for any differences from your predictions. (Hint: this is a challenging exercise; you should be sure to be able to reproduce the results in Example 2 before considering these modifications. You may have to "force" the model to fit the situation you want to apply it to.)

a)   Increasing the yield factor of HCN to 50% from the combustion process.

[Note: relevant computer output files are included at the end of this solution set.] The toxic hazard distance increases from $1,977'$ to $5,890'$. The evolution rate of HCN is increased as the model source term from 29 lb/min to 145 lb/min, and the toxic vapor dispersion calculations yield this increased distance.

b)   Assuming ignition takes place only after the pool reaches maximum size.

This is the hardest part of the problem. It requires a second toxic vapor dispersion model run for HCN, corresponding to ignition of the pool after it reaches a maximum area of about 35,790 ft². To get the source term for this release, we need to know how fast the acrylonitrile would be burning in this case, but our model does not provide this as an output. However, the rate of burning is roughly proportional to the surface area of the pool (since any given portion of the material just burns from the top of the pool down without being affected by what is going on on "each side" of it), and we do indeed know the burn rate of acrylonitrile from parts b and d in the text discussion of the HCN modeling, 569.2 lb/min, as well as the corresponding burning pool area, 1,186 ft², calculated by the model in that case. Therefore, our larger pool, which is larger by a factor of $(35,790/1,186) = 30.2$, should burn at a total rate about 30.2 times larger than the earlier value of 569.2. When we apply this factor to the earlier calculated HCN production rate of 29 lb/min, we get a new HCN source rate of about 875 lb/min. This would be released at roughly the flame height of $155'$ calculated in part a in the text discussion.

Note that this modified situation is very short-lived, as there is only enough mass to burn at this rate for about 3 minutes, so this is basically a puff release. The downwind hazard distance is calculated to be $10633'$. (There is some additional conservatism here in that the model treats the large burning area as a point source.)

c)   Changing the weather conditions to moderate sunshine.

Changing the conditions to moderate sunshine corresponds to changing the stability class from D to B. The new hazard distance for HCN (under the other baseline assumptions) is found to be 744′.

9.   Under what weather conditions would the toxic hazard zone from acrylonitrile vapors in Example 2 extend the farthest distance from the scene? What about the case when the accident could also occur at night? (Hint: these two questions may require some experimentation on your part; keep your results well organized and present them in tabular form.) Give a plausible physical explanation for your conclusion.

This requires the repetitive application of the toxic vapor model for HCN with a source strength of 29 lb/min, varying only the wind speed and stability class. Table 4-F, which goes beyond the specific questions asked in this exercise, reflects a range of wind speeds for each stability class. It can be seen that for daytime conditions, D stability (overcast conditions) with the lowest possible windspeed for this class will lead to the maximum hazard distance. The ARCHIE program restricts the wind range for D stability to 4.5 mph or above, and this gives the largest daytime value shown in the table (2,139 feet). Allowing for the incident to occur during the night, F stability (relatively clear conditions) with the lowest possible wind speed for this class, 4.5 mph, yields the longest hazard distance as 5,038 feet. (The restriction of wind speeds to be above certain minimum values is usually based on the relatively rare occurrence of lower values. Even when the air might feel very still at ground level, there is usually still some small wind, which is also greater as one moves slightly higher. Few people have a well-calibrated internal

TABLE 4-F
Numerical results relating hazard distance to stability class and wind speed

Stability class	Wind speed	Toxic hazard distance	Peak location	Peak value
A	1	1186	245	251.8
	2	871	252	125.9
	5	529	242	50.4
B	1	1844	341	266.8
	2	1271	331	133.3
	5	744	340	53.4
	7	588	345	38.1
	11	none	347	24.3
C	5	1165	479	58.2
	10	663	485	29.1
	15	none	486	19.4
D	4.5	2139	787	59.8
	5	1977	802	53.8
	7	1503	798	38.5
	11	none	789	24.5
E	5	2855	1152	51
	7	2119	1140	36.4
	11	none	1123	23.2
F	4.5	5038	1935	49.1
	5	4577	1959	44.2
	6	3829	1971	36.8

sense of the wind speed.) In both of these cases, the most stable class (which tends to keep the plume together) and lowest corresponding wind speed (which minimizes dilution) lead to the maximum hazard distances.

Table 4-F also provides additional opportunities to track trends and confirm them against your physical intuition and your understanding of stability classes. For example, within any stability class, the hazard distance is greater as the windspeed decreases, primarily because of the higher dilution provided by increased wind. Similarly, within a given class, the peak location stays at about the same point as the wind speed varies, although once again the value of the peak would decrease due to dilution. These kinds of relationships are good for students to investigate and discuss.

To facilitate the verification of the values in the text for Example 2 and for the above variations, here are six abbreviated ARCHIE outputs, the first three corresponding to items a, b, and d in the text, and the next three corresponding to the three parts of Exercise 8. However, it is strongly recommended that teachers themselves work out any problems that they intend to assign or discuss with the students, as there may be a number of decisions that need to be made along the way, and the teacher should be aware of the kinds of issues that are likely to cause problems. Therefore these printouts are of greatest utility in helping teachers debug their own solutions if difficulties arise.

Text "part a":

```
HAZARDOUS MATERIAL    = acrylonitrile
NAME OF DISK FILE     = BOOKACR1.ASF


 ***  SCENARIO DESCRIPTION
      Basic analysis to hal idlh and lfl for
      chemical itself. Spread on water max size.


 ******  DISCHARGE RATE/DURATION ESTIMATES
      Liquid discharge from nonpressurized container
      Average discharge rate    = 569.2      lbs/min
      Duration of discharge     = 95         minutes
      Amount discharged         = 54050      lbs
      State of material         = Liquid

 ******  LIQUID POOL SIZE ESTIMATES
      Evaporating pool area     = 35790      ft2
      Burning pool area         = 35889.9    ft2

 ******  LIQUID POOL EVAPORATION RATE/DURATION ESTIMATES
      Vapor evolution rate      = 569.2      lbs/min
      Evolution duration        = 95         minutes

 ******  TOXIC VAPOR DISPERSION ANALYSIS RESULTS
      Downwind distance to concentration of 250 ppm
      -- at groundlevel         = 2288       feet

 ******  POOL FIRE HAZARD ESTIMATION RESULTS
      Burning pool radius       = 106.8      feet
      Flame height              = 155        feet
      Fatality zone radius      = 341        feet
      Injury zone radius        = 488        feet
```

```
******* FLAMMABLE VAPOR CLOUD HAZARD RESULTS
            For concentration of        1/2 LFL    LFL
                                         ------     -----
            Downwind hazard distance   =  230       159      feet
            Max hazard zone width      =  115        80      feet
            Max weight explosive gas   =  298       206      lbs
            Relative gas/air density   =  1.14      1.14     initially
            Model used in analysis     =  Neutrally buoyant
   Note: Clouds or plumes containing less than 1000 pounds of vapor or gas
         are very unlikely to explode when completely unconfined, except
         when one of a certain few materials have been discharged.


                    INPUT PARAMETER SUMMARY
         ----------------------------------------
PHYSIOCHEMICAL PROPERTIES OF MATERIAL
    NORMAL BOILING POINT            = 172        degrees F
    MOLECULAR WEIGHT                = 53.06
    LIQUID SPECIFIC GRAVITY         = .81
    VAPOR PRES AT CONTAINER TEMP    = 2.372      psia
                                    = 122.8      mm Hg
    VAPOR PRES AT AMBIENT TEMP      = 2.38       psia
                                    = 122.8      mm Hg
    LOWER FLAMMABLE LIMIT (LFL)     = 3          vol%
    TOXIC VAPOR LIMIT               = 250        ppm
CONTAINER CHARACTERISTICS
    CONTAINER TYPE                  = Horizontal cylinder
    TANK DIAMETER                   = 7          feet
    TANK LENGTH                     = 40         feet
    TOTAL WEIGHT OF CONTENTS        = 54050      lbs
    WEIGHT OF LIQUID                = 54050      lbs
    LIQUID HEIGHT IN CONTAINER      = 4.6        feet
    WEIGHT OF GAS UNDER PRESSURE    = 0          lbs
    TOTAL CONTAINER VOLUME          = 1539       ft3
                                    = 11513      gals
    LIQUID VOLUME IN CONTAINER      = 1069       ft3
                                    = 7997       gals
    DISCHARGE HOLE DIAMETER         = 2.26       inch(es)
    DISCHARGE COEFFICIENT OF HOLE   = .62
    TEMP OF CONTAINER CONTENTS      = 82         degrees F
ENVIRONMENTAL/LOCATION CHARACTERISTICS
    AMBIENT TEMPERATURE             = 82         degrees F
    WIND VELOCITY                   = 5          mph
    ATMOSPHERIC STABILITY CLASS     = D
    LIQUID CONFINEMENT AREA         = NONE
    VAPOR/GAS DISCHARGE HEIGHT      = 0          feet
KEY RESULTS PROVIDED BY USER INSTEAD OF BY EVALUATION METHODS
    NONE OBSERVED

KEY RESULTS OVERRIDDEN BY USER AT SOME POINT AFTER COMPUTATION
    NONE OBSERVED
```

Text "part b":

```
HAZARDOUS MATERIAL    = acrylonitrile
NAME OF DISK FILE     = BOOKACR2.ASF

*** SCENARIO DESCRIPTION
    Only to get flame data for HCN model, assuming
    burning at rate of discharge.

******* DISCHARGE RATE/DURATION ESTIMATES
        Liquid discharge from nonpressurized container
        Average discharge rate   = 569.2        lbs/min
        Duration of discharge     = 95           minutes
        Amount discharged         = 54050        lbs
        State of material         = Liquid

******* LIQUID POOL SIZE ESTIMATES
        Evaporating pool area     = 35790        ft2
        Burning pool area         = 1186         ft2

******* POOL FIRE HAZARD ESTIMATION RESULTS
        Burning pool radius       = 20           feet
        Flame height              = 48           feet
        Fatality zone radius      = 62           feet
        Injury zone radius        = 89           feet
```

Text "part d":

```
Hazardous Material    = hcn
Name of disk file     = BOOKHCN2.ASF

*****   SCENARIO DESCRIPTION
        HCN source based on 10% yield factor

*****   TOXIC VAPOR DISPERSION ANALYSIS RESULTS
        Downwind distance to concentration of 25 ppm
        -- at groundlevel         = 1977   feet
        -- at discharge height    = 1786   feet

        Peak concentration on ground is 53.8 ppm
        at a downwind distance of 802 feet for
        elevated emission source specified by user.

        User changed vapor emission rate prior to use of
        toxic or flammable vapor dispersion model.
        Final user provided rate was 29 lbs/min.

        User changed vapor emission duration prior to use
        of toxic or flammable vapor dispersion model. Final
        user provided duration was 95 minutes.
```

Exercise 8a output:

```
Hazardous Material    = hcn
Name of disk file     = EX8AS4-6.ASF

*****   SCENARIO DESCRIPTION
        example 2 but with hcn increased to 50% yield,
        hence 5X original value in BOOKHCN2.
```

***** TOXIC VAPOR DISPERSION ANALYSIS RESULTS

Downwind distance to concentration of 25 ppm
-- at groundlevel   = 5890   feet
-- at discharge height = 5714   feet

Peak concentration on ground is 269.1 ppm
at a downwind distance of 790 feet for
elevated emission source specified by user.

User changed vapor emission rate prior to use of
toxic or flammable vapor dispersion model.
Final user provided rate was 145 lbs/min.

User changed vapor emission duration prior to use
of toxic or flammable vapor dispersion model. Final
user provided duration was 95 minutes.

Exercise 8b output:

Hazardous Material   = hcn
Name of disk file   = EX8BS4-6.ASF

***** SCENARIO DESCRIPTION
hcn output from full size pool fire and flame
height. Short duration since mass used quickly.

***** TOXIC VAPOR DISPERSION ANALYSIS RESULTS

Downwind distance to concentration of 25 ppm
-- at groundlevel   = 10633 feet
-- at discharge height = 9965   feet

Peak concentration on ground is 117.9 ppm
at a downwind distance of 3130 feet for
elevated emission source specified by user.

User changed vapor emission rate prior to use of
toxic or flammable vapor dispersion model.
Final user provided rate was 875 lbs/min.

User changed vapor emission duration prior to use
of toxic or flammable vapor dispersion model. Final
user provided duration was 2.06 minutes.

Exercise 8c output:

Hazardous Material   = hcn
Name of disk file   = EX8CS4-6.ASF

***** SCENARIO DESCRIPTION
changing weather conditons to moderate sunshine
***** TOXIC VAPOR DISPERSION ANALYSIS RESULTS

Downwind distance to concentration of 25 ppm
-- at groundlevel   = 744   feet
-- at discharge height = 683   feet

Peak concentration on ground is 53.4 ppm
at a downwind distance of 340 feet for

```
elevated emission source specified by user.

User changed vapor emission rate prior to use of
toxic or flammable vapor dispersion model.
Final user provided rate was 29 lbs/min.

User changed vapor emission duration prior to use
of toxic or flammable vapor dispersion model. Final
user provided duration was 95 minutes.
```

In finishing the above problems, with all their various complications, you might be wondering how you can make up your own problems for exams or problem sets. You don't need to be a chemist to do this. If you really want to make up realistic problems, there are at least three good ways to pick a chemical and get ideas for scenarios. First, you could read some newspaper accounts of accident reports. Try using the same chemical to define an accident scenario, either similar to the one you read about or of your own choosing. If you access the chemical's MSDS on-line or otherwise, you will quickly be able to see whether it has enough flammability or toxicity hazard to be "interesting." Second, you could look through the annual reports submitted by local companies to the LEPC or fire department concerning their inventories of hazardous chemicals. This is public information and is sometimes available in the local library as well. A scenario involving a hazardous chemical in heavy local use would probably be inherently interesting. Third, you can access a variety of government databases, such as through the EPA or environmental organizations, that give data on reportable toxic releases to the environment. From this you will also get ideas about both chemicals and scenarios.

4.7 General Comments and Guide to Further Information

(No exercises.)

5

Additional Topics in Ground Water

Many of the problems in this section involve numerical calculations that can become tedious if done by hand. Although students are probably already accustomed to using spreadsheets or other programs to perform many of their calculations, they should perhaps once again be encouraged to set up neat, versatile programs that they can adapt from one problem to another. In addition, rather than continuing to modify and save the spreadsheet or other program under the same name, it can be useful to save it each time under a name that corresponds to the problem for which it is being applied. In this way, one can always go back and access the calculations for a given problem. Furthermore, since some later problems involve modifications to earlier ones, this provides a complete record of each problem for later enhancement. This is also a good way for the teacher to keep track of the problems and to generate modifications for additional practice or testing.

It can also be valuable in class to discuss some of the computational shortcuts some students may have worked out. There is usually quite a wide range in their ability to handle such computations efficiently, and it is more effective if they hear about short cuts from their peers. I would encourage them to use the computer for graphing and for all the computations possible. In this way, it will become second nature to them and they will not be intimidated by problems that don't work out evenly or that have numbers of strange sizes.

5.1 The Continuous Form of Darcy's Law for One-Dimensional Flow

1. **Suppose on the basis of some data collection and curve fitting, a hydrologist produces the following equation to represent the approximate hydraulic head value as a function of distance x in an aquifer, measured in an easterly direction from a fixed location:**

$$h(x) = 29 + 0.02x - 0.00000063x^2.$$

This is to believed to be an acceptable approximation on the interval from $x = 0$ to $x = 12,000$ feet.
a) **Is the flow from east to west or west to east?**

Both the head function and its derivative are graphed in Figure 5-A. Since the head is larger to the right (east), the flow would be from east to west.

b) **What is the value of the hydraulic gradient at the point corresponding to $x = 6,000$?**

This would be the derivative of the head function at that point, which is:

$$h'(6000) = .02 - 2(0.00000063)(6000) = 0.0124.$$

So we would generally say that the gradient is 0.0124 from east to west. (It would not generally be reported to this precision, but it is useful here just to check that the right computation has been performed.

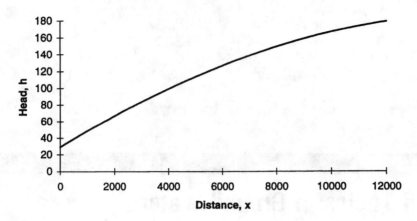

FIGURE 5-A
Head function for Exercise 1

c)   Where is the hydraulic gradient the greatest along the given interval?

At the left end of the interval, since the curve is steepest there. (Alternatively, you could look at the derivative function and see that it is largest at the left end.)

2.   Similarly to the situation in the previous exercise, the hydraulic head in a different aquifer is modeled by the equation:

$$h(x) = 112 + 0.03x - 0.00000028x^2,$$

although in this case the x-direction is measured north from a fixed point. Identify what might be considered to be peculiar behavior in this aquifer and describe geologic conditions that would be a possible explanation. Can you think of a topographic analog involving surface water flow?

The graph of this head function is shown in Figure 5-B. Since it has a peak in the center portion of the interval, there is a "ground-water divide," similar to the effect one encounters on topographic ridges or rises, where the surface water drains in opposite directions on both sides. In this case, one possible explanation might be that the center is the high point of an upwarped water table aquifer formation.

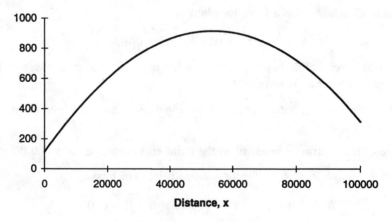

FIGURE 5-B
Head function for Exercise 2

3. **Returning to the aquifer in Exercise 1, suppose that it is uniformly 20 feet thick, 700 feet wide in the north-south direction, and has a hydraulic conductivity of 6.2 ft/day. Why would this situation be impossible?**

Every cross section of the aquifer perpendicular to the flow axis must carry water at the same total volumetric flow rate. This is just conservation of mass: everything that comes out of one portion has to go into the next at the same rate. The total flow rate is

$$Q = qA = -K\frac{dh}{dx}A = -6.2 \times \frac{dh}{dx} \times 20 \times 700 = -86800 \times \frac{dh}{dx}$$

so this could be constant from one location x to another only if the derivative $\frac{dh}{dx}$ were constant, which it clearly is not.

5.2 Applying the Continuous One-Dimensional Version of Darcy's Law

1. **Assume in the context of the above discussion that we actually do measure a value $h(2000) = 65$. Assume further that we decide to assume a constant value of hydraulic conductivity to the left of $x = 2,000$ and a possibly different constant value of hydraulic conductivity to the right of $x = 2,000$. Without doing any computations but just thinking about the physical nature of this problem, would you expect the value of hydraulic conductivity on the left to be greater than or less than the value on the right? Explain your reasoning.**

We would expect the value on the left to be greater than the value on the right. Since the left side will have a smaller hydraulic gradient (which encourages less flow), it must compensate for this by having a greater hydraulic conductivity, in order for the flow rates over the two sections to be equal, which they must be by conservation of mass. Another way to think about this is that the anomalously high value of 65 at the midpoint suggests that the water is getting "backed up" on the right side, just as when you dam a river or clog a drain, the water backs up to a higher level. This would mean a lower conductivity value on the right.

2. **Continuing the above discussion with the condition $h(2000) = 65$, can you determine specific values for K_{left} and K_{right} that must exist underground in order to lead to this distribution of head values? If so, find these values. If not, find the most detailed relationship you can between these values.**

By equating flows on the left and right sections, we obtain:

$$K_{\text{left}} i_{\text{left}} A = K_{\text{right}} i_{\text{right}} A$$

$$K_{\text{left}} i_{\text{left}} = K_{\text{right}} i_{\text{right}}$$

$$K_{\text{left}} \frac{70 - 65}{1000} = K_{\text{right}} \frac{65 - 50}{1000}$$

$$K_{\text{left}}(.005) = K_{\text{right}}(.015)$$

$$K_{\text{left}} = 3K_{\text{right}}.$$

This is the most precise or detailed relationship implied by the given information.

3. **Figure 5-5 shows a section view similar to a proposed nuclear waste repository site located deep beneath the surface in the southwestern United States. (A repository is a mine in which such wastes would be placed and then sealed off permanently.) The aquifer shown on the diagram is a dolomite bed that averages 25 feet thick, but whose characteristics gradually change to the west (i.e., towards the left side of the diagram). In particular, between Q and R, the average hydraulic conductivity is 0.03 ft/day with a porosity of .05, whereas from P to Q, the hydraulic conductivity increases to 0.3 ft/day due to increased fractures and solutioning, and the porosity increases to .10. The heads at P and R were measured at test wells and found to be 560**

and 685 feet, respectively. Assume that the repository and the affected portion of the aquifer directly above it are 400 feet wide in the north-south direction (i.e., perpendicular to the section shown above), and use the scale on the diagram to determine distances.

a) What must the head be at Q?

We begin by estimating the distances using the map scale. The distance from P to Q is 2.5 miles (13200 ft), and the distance from Q to R is 1.5 miles (7920 ft). We now balance the flow rates on the two sections, which must be equal by the usual conservation principle:

$$.3 \times \frac{h_Q - 560}{13200} \times A = .03 \times \frac{685 - h_Q}{7920} \times A$$

$$h_Q - 560 = \frac{13200}{.3} \times \frac{.03}{7920} \times (685 - h_Q)$$

$$h_Q = 560 + 114.167 - \frac{h_Q}{6}$$

$$h_Q = \frac{6}{7} \times (674.167) = 577.86.$$

Thus the final (rounded) head value at Q would be 578 ft. (Note that the A's canceled without our having had to input their precise common value.)

b) What would be the ground-water travel time from point R to point P, just in case some radioactive material leaked from the repository and mixed with the water in the aquifer?

The following table gives all the intermediate terms used in applying the interstitial velocity equation to calculate travel times.

Segment	h(right)	h(left)	length	gradient	K	porosity	velocity	time
QR	685	577.857	7920	0.013528	0.03	0.05	0.008117	975742.7
PQ	577.857	560	13200	0.001353	0.3	0.1	0.004058	3252506
TOTAL (days)								4228249
TOTAL (years)								11584.24

Thus the total travel time would be about 11,600 years.

c) Find an equation for the head function $h(x)$ where x is distance measured west from point R.

The hydraulic gradient must be constant for the flow to be the same through every cross section within each of the two portions of the aquifer where the other properties are constant. Thus the slope of the head function must be constant, and hence the head function itself must be a straight line. Since we know its endpoint values on each of these intervals, we simply need to find the equation of the straight line passing through these points.

As x ranges from 0 (point R) to 7920 (point Q), $h(x)$ is a straight line passing through the points $(0, 685)$ and $(7920, 577.857)$. Thus the slope is given by

$$m = \frac{577.857 - 685}{7920 - 0} = -0.0135,$$

and the y-intercept is just 7920. Therefore, on this interval, the line is given by

$$h = -.0135x + 7920.$$

Similarly, on the left interval the points are $(7920, 577.857)$ and $(21120, 560)$, so that the slope is

$$m = \frac{560 - 577.857}{21,120 - 7,920} = -0.00135$$

which yields a y-intercept of 588.57, and hence the equation

$$h = -.00135x + 588.57.$$

Summarizing, the final combined equation for h is:

$$h = \begin{cases} -.0135x + 7,920 & \text{for } 0 \le x \le 7,920 \\ -.00135x + 588.57 & \text{for } 7,920 \le x \le 21,120 \end{cases}$$

d) Where is the hydraulic gradient the largest?

It is greatest on the right portion of the flow pathway, as can be seen by looking at the slope or derivative of the two equations immediately above. (Remember, we generally talk about the hydraulic gradient as being the corresponding positive value, so the larger gradient would be the value 0.0135 experienced on the right side.)

e) Is all of the given quantitative information necessary in order to answer the above questions?

We did not need the precise values of the width or thickness of the aquifer, but it was essential to know that the cross section remained constant throughout the entire distance. Otherwise, we would not have been able to cancel out the A values in part a.

4. Describe the complete set of solutions to the one-dimensional Laplace equation. Be sure to explain why no other function, no matter how strange, could possibly be a solution.

The set of all (non-vertical) straight lines. The standard of rigor for the explanation certainly depends on the mathematical background of the student or reader and what you wish to assume from calculus, but one could argue along the following lines for a mathematical proof.

Suppose $h(x)$ is any function of x that is a solution to the equation $h_{xx} = 0$, say even on some limited interval within which we pick a fixed point a. Integrating both sides of the differential equation from a to an arbitrary point x, we obtain:

$$\int_a^x h_{xx}dx = \int_a^x 0\,dx$$

$$h'(x) - h'(a) = 0 \quad \text{(changing to the familiar prime notation for derivatives)}$$

$$h'(x) = C \quad \text{(for a constant } C \text{ that we use to stand for } h'(a))$$

$$\int_a^x h'(x)\,dx = \int_a^x C\,dx$$

$$h(x) - h(a) = C(x - a)$$

$$h(x) = Cx + D \quad \text{(for the obvious constant } D = h(a) - Ca).$$

From this last equation, we conclude that h must indeed be a simple linear function.

5. Where would the derivation of the one-dimensional Laplace equation fail if the hydraulic conductivity were not constant, but rather were to vary along the aquifer?

The proof would fail in the following mass balance step:

$$0 = K\big(-h_x(x)\big)(1 \cdot 1) + K\big(h_x(x) + \Delta x\big)(1 \cdot 1)$$

because we would not have the same single K value for both terms. This would then make it impossible to cancel the K's and proceed with the limiting process.

6. Suppose that in the same context as above we have an aquifer within which we hypothesize that the hydraulic conductivity varies linearly with x. That is, we assume that the hydraulic conductivity can be represented by an equation of the form

$$K = K_0 + \alpha x$$

for some constant α. What differential equation would h satisfy in this case?

In this case the mass balance would be written as:

$$0 = \text{flow in through left face} + \text{flow in through right face}$$

$$0 = (K_0 + \alpha x)(-h_x(x))(1 \cdot 1) + (K_0 + \alpha(x + \Delta x))(h_x(x + \Delta x))(1 \cdot 1).$$

The subsequent terms would thus be slightly more complicated than before:

$$0 = \text{flow in through left face} + \text{flow in through right face}$$

$$0 = (K_0 + \alpha x)(-h_x(x))(1 \cdot 1) + (K_0 + \alpha(x + \Delta x))(h_x(x + \Delta x))(1 \cdot 1)$$

$$0 = (K_0 + \alpha x)\frac{h_x(x + \Delta x) - h_x(x)}{\Delta x} + \alpha h_x(x + \Delta x)$$

$$0 = \lim_{\Delta x \to 0} \frac{h_x(x + \Delta x) - h_x(x)}{\Delta x} + \lim_{\Delta x \to 0} \frac{\alpha h_x(x + \Delta x)}{(K_0 + \alpha x)}$$

$$0 = h_{xx}(x) + \frac{\alpha h_x(x)}{(K_0 + \alpha x)}.$$

(You can see that if $\alpha = 0$, this reduces to the case of Laplace's equation, as expected.)

7.   Return to the situation of Figure 5-4 with the observation that $h(2000) = 65$ and the assumption that there are constant values of hydraulic conductivity on the intervals both to the left and to the right of the point $x = 2,000$. Draw a graph of the function $h(x)$ that would be the solution to this problem on the interval from $x = 1,000$ to $x = 3,000$.

This is just a polygonal line passing through the data points, as shown in Figure 5-C.

8.   Figure 5-7 shows a still more detailed structure for a one-dimensional flow problem. In this case the interval from the 1,000-ft marker to the 3,000-ft marker is actually divided into three parts, and the head values are measured at all the nodes. These values are shown on the figure. The K values are assumed constant on each of the three sub-intervals. Furthermore, the hydraulic conductivity within the leftmost sub-interval has been extensively investigated and is known to have the value of 0.4 ft/day.

a)   Find the hydraulic conductivity in the other two sections of the aquifer.

The value of the product Ki must be the same in each of the sub-intervals because of mass conservation. In the leftmost interval, we have enough information to calculate this product:

$$Ki = .4 \times \frac{70 - 68}{600} = 0.001333.$$

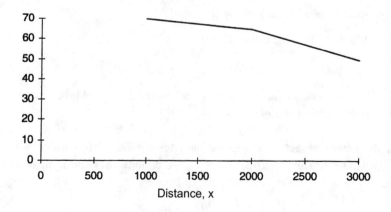

FIGURE 5-C
Graph for Exercise 7

Therefore K for the middle interval can be found from the equations

$$Ki = .001333$$

$$K\left(\frac{68 - 60}{900}\right) = .001333$$

$$K = 112.5 \times .001333 = .15 \text{ ft/day}.$$

Similarly, for the rightmost interval, we obtain

$$Ki = .001333$$

$$K\left(\frac{60 - 50}{500}\right) = .001333$$

$$K = 50 \times .001333 = .067 \text{ ft/day}.$$

b)   Find the head value at the point $x = 2{,}800$.

The head function on the rightmost interval will be the linear interpolant to the endpoints $(2500, 60)$ and $(3000, 50)$. This straight line has slope

$$m = \frac{50 - 60}{500} = -.02,$$

and the y-intercept is found to be 110. Hence the equation is

$$h = -.02x + 110.$$

Plugging 2,800 into this equation, we obtain $h(2800) = 54$. (This result can also be determined by the principle of proportion, in that 2800 is three-fifths of the way from the endpoint 2,500 to the other endpoint 3,000, so the head value should be three-fifths of the way from 60 to 50, which gives 54.)

c)   Write an equation that will give the head value at any point in the interval.

While this could be worked out using the detailed method of part b, let us for convenience apply the principle of proportion as illustrated at the end of part b. This enables us to write down all the answers by inspection:

$$h(x) = \begin{cases} 70 - 2\left(\dfrac{x - 1000}{1600 - 1000}\right) & \text{for } 1000 \leq x \leq 1600 \\[2mm] 68 - 8\left(\dfrac{x - 1600}{2500 - 1600}\right) & \text{for } 1600 \leq x \leq 2500 \\[2mm] 60 - 10\left(\dfrac{x - 2500}{3000 - 2500}\right) & \text{for } 2500 \leq x \leq 3000. \end{cases}$$

d)   If the porosity values of the three subintervals are .35, .3, and .2, from left to right, how long would it take ground water to travel from $x = 1{,}000$ to $x = 3{,}000$?

A tabular summary of these calculations is given below:

Segment	h(left)	h(right)	length	gradient	K	porosity	velocity	time
left	70	68	600	0.003333	0.4	0.35	0.00381	157500
middle	68	60	900	0.008889	0.15	0.3	0.004444	202500
right	60	50	500	0.02	0.066667	0.2	0.006667	75000
TOTAL (days)								435000
TOTAL (years)								1192

Thus the travel time would be almost 1,200 years.

9.   Figure 5-8 is a revised interpretation of the nuclear waste repository setting treated earlier in this section. However, the aquifer in this case is now hypothesized to be divided into three distinct sections, with parameter values as shown on the figure. Head values are known only at the extreme endpoints, as previously.

a)   If you were to drill monitoring wells at points S and T, what head values would be predicted by this conceptual model of the aquifer?

We begin by using the scale to determine the distances: $\overline{PS} = 3520$, $\overline{ST} = 11440$, and $\overline{TR} = 6160$. In this case we need to find two unknown head values, and we will do this by setting up two simultaneous equations in these two unknowns. Once again, the basic principle is that of mass conservation. The first equation balances the flows at S:

$$.5\left(\frac{h_S - 560}{3520}\right) = .07\left(\frac{h_T - h_S}{11440}\right),$$

and the second equation balances the flows at T:

$$.07\left(\frac{h_T - h_S}{11440}\right) = .002\left(\frac{685 - h_T}{6160}\right).$$

These two equations reduce to the pair

$$1.043h_S - .043h_T = 560$$

$$-h_S + 1.053h_T = 36.347$$

whose solutions are

$$h_S = 560.3$$

$$h_T = 566.6.$$

(Spreadsheet programs can generally be used to invert matrices or solve linear equations.) These are the values that would be predicted by this model. You can see that since they are so close to the left value of 560, this means that almost all the drop in head occurs over the segment $\overline{TR}$, which implies that this is the segment that provides the greatest resistance to flow. A review of the K values confirms that this should be the case, as this is the segment with the lowest hydraulic conductivity.

b)   What would be the ground-water travel time from R to P according to this model?

From a computational standpoint, this is now identical to Exercise 8, so we simply adapt the same spreadsheet program. The results are shown in the table below.

Segment	h(right)	h(left)	length	gradient	K	porosity	velocity	time
left	560.3	560	3520	8.52E-05	0.5	0.2	0.000213	1.65E+07
middle	566.3	560.3	11440	0.000524	0.07	0.1	0.000367	3.12E+07
right	685	566.3	6160	0.019269	0.002	0.008	0.004817	1.28E+06
TOTAL (days)								4.90E+07
TOTAL (years)								134136

An interesting observation here is that even though the rightmost segment provides by far the greatest resistance to flow, the travel time for the middle segment is the largest. This is because the lower porosity for the right segment has a counterbalancing effect when it enters the denominator of the interstitial velocity equation.

10.   For the situation shown in Figure 5-9, assume the following parameter values. The confined aquifer is 20 feet thick and is fully penetrated and screened by a well that is 8 inches in diameter. Its hydraulic conductivity is 28 ft/day. During pumping, the head at the well is maintained at 44 feet, whereas the head at the limit of the radius of influence (estimated to be 50 feet from the well) is 55 feet. Find the head function $h(r)$ and the quantity of water (in gallons per minute) being extracted from the well under these conditions. How sensitive is this latter value to the assumed radius of influence of the well?

Imagine each contour line in the figure as denoting a "vertical curtain" in the aquifer through which water is flowing perpendicularly on its way towards the well. The area of the curtain would be the surface area of a cylinder, or $2\pi r(20)$ in this case because the aquifer is 20 feet thick. By conservation of mass, the flow through each such cylindrical surface must be the same. The hydraulic gradient would just be $-\frac{dh}{dr}$, so the flow through any such surface can be gotten by Darcy's law:

$$Q = -K\frac{dh}{dr}(40\pi r) = \text{some constant } C.$$

C is a constant; it is independent of r since the flow must be the same for cylindrical surfaces of any radius r. Now we integrate this equation with respect to r, and there are two somewhat different ways to do this. We could take an indefinite integral of each side and add a constant of integration, and then use the boundary conditions at the well radius and at the radius of influence to solve two simultaneous equations for C and for this new constant of integration. Alternatively, we can take the definite integral of each side from the inner radius (4 inches, or a third of a foot) to an arbitrary point r:

$$\frac{dh}{dr} = \frac{C}{40\pi K r}$$

$$\int_{1/3}^{r} \frac{dh}{dr}\,dr = \int_{1/3}^{r} \frac{C}{40\pi K r}\,dr$$

$$h(r) - h(\tfrac{1}{3}) = \frac{C}{3518.6}\ln(3r)$$

$$C = \frac{3518.6(55-44)}{\ln(3 \times 50)} = 7724.5.$$

$$h(r) = 44 + 2.2\ln(3r)$$

This completes the first part of the problem. Now we apply any convenient radius (say, $r = 1$) to find the flow through each "curtain" or cylindrical surface, because that must be the same as the amount of water being extracted from the well:

$$Q = -K\frac{dh}{dr}A = -28 \times 2.2 \times 40\pi = 7741 \text{ cubic feet per day.}$$

This converts to about 40 gallons per minute.

Now we discuss the sensitivity of this result to the assumed radius of influence, since the latter is usually just a rough estimate, not a precise, measurable value. From the above equations it is clear that the flow rate is proportional to the constant C and that the radius of influence, called R, enters into the determination of C in a denominator through a factor $\ln\left(\dfrac{R}{r_{\text{well}}}\right)$. Therefore, if you were to double the radius of influence, this denominator would change only by an additive factor of $\ln 2$, due to the laws of logarithms. Even if you multiplied your estimate of the radius of influence by 10, the denominator would be increased only by an additive factor of $\ln 10$, hardly causing any change! What is going on here? The answer is that the overwhelming control on the amount of water entering the well is due to the relatively small neighborhood around the well where the flow lines converge, and it is harder for the water to have enough cross section to move through. So there is a "back-up" in this area, the hydraulic gradient gets steep there (as can be verified from the logarithmic equation for h), and there is not much influence at all from what is going on at a greater distance away from the well.

5.3 Retardation Factors

1. **This problem is an extension of Exercise 3 in Section 5.2, to which you should refer for the basic figure and parameter values. You should have already calculated as part of that exercise the ground-water travel time from R to P, expressed in years. If a radioactive substance were to enter the ground water through a leak from the repository at R, how long would it take to get to P, assuming its retardation factor in the RQ portion of the aquifer is 50 and its retardation factor in the QP portion of the aquifer is 10?**

This simply involves reducing the velocity in each segment by the corresponding retardation factor, or, equivalently, increasing the travel time in the segment by the same factor. The results from the cited exercise are modified below to incorporate these changes.

Segment	water time	time with retardation
QR	975,743	48,787,135
PQ	3,252,506	32,525,060
TOTAL (days)	4,228,249	81,312,195
TOTAL (years)	11,584	222,773

Thus the resulting travel time would be almost 223,000 years.

You may wonder whether travel times of this magnitude are actually calculated in real applications. Yes, they are—primarily in applications involving the estimated performance of radioactive waste disposal systems. Naturally there are large uncertainties in calculations such as this, but the order of magnitude of the results does give an indication of whether such substances could reach the so-called "accessible biosphere" or "accessible environment" before decaying to harmless levels by radioactive processes.

2.   Consider a system in which the bulk rock density is 2.3 g/cm^3 and where the porosity is 30%. In order for the retardation factor to turn out to be 100, what would the K_d value have to be? (Use the relationship between the retardation factor and the distribution coefficient given in the text.) Furthermore, make a simple statement about the percent of mass of solute that would be adsorbed onto the solid phase in this system (analogous to the hypothetical 99% figure that was used in the text).

Substituting into the relationship between K_d and the retardation R, we have:

$$R = 1 + \frac{\rho_b}{\eta} \times K_d$$

$$100 = 1 + \frac{2.3 \text{ g/cm}^3}{.3} \times K_d$$

$$K_d = \frac{30 - .3}{2.3 \text{ g/cm}^3} = 12.91 \text{ cm}^3/\text{g} = 12.91 \text{ ml/g}$$

where we have replaced the cubic centimeter unit by milliliters, which are the same thing but just more common for this particular parameter.

To answer the question about the percent adsorbed, suppose we had 1 milliliter of solution of initial concentration C, which would thus contain C mass units of solute. (We need to carry these calculations out with a high degree of precision, or express them as fractions, so that the rounding process does not hide differences in results. For this purpose, the K_d value from above is actually 12.913043.) If we poured this milliliter of solution onto the rock, it would then fill up 1 cm^3 of pore volume. Since the porosity of the bulk rock is 30%, this pore volume would correspond to a bulk rock volume of $1/.3 = 3.3333333$ cm^3, and hence to a rock mass of $2.3 \times 3.3333333 = 7.6666667$ g. Much of the material would be adsorbed, leaving a new equilibrium concentration D in the liquid phase (and since we have 1 cm^3 of solution, D has the same numerical value as the residual mass in solution). From the definition of K_d, each gram of affected rock will adsorb $12.913043D$ units of solute, for a total of $12.913043D \times 7.6666667 = 99D$ g. The relation between C and D is that

$$99D + D = C$$

$$D = \frac{C}{100} = .01C.$$

Thus the fraction of material that is adsorbed will be:

$$\frac{99D}{C} = \frac{99(.01C)}{C} = .99 = 99\%,$$

which for this particular case happens to be the same as the hypothetical value used in the text.

3.   Derive the equation given in the text for the retardation factor as a function of the distribution coefficient.

Suppose we have solution of constant concentration C beginning to enter a segment of the flow path and that the ground water itself is moving with velocity v. Following the description of Figure 5-11 in the text, each short segment of length d would become fully saturated with adsorbed solute before the concentration front moved downstream to the next segment. (This is a consequence of our assumption of an instantaneous equilibrium reaction. Departures from this assumption in real cases simply imply that some solute will make it through the system at an earlier time, although generally at a very low concentration.) Consider the first such segment of length d. The maximum amount of solute that can be forced into the adsorbed state by the adsorption equilibrium reaction is the amount that corresponds to the input concentration C. By the definition of K_d, this amount is

$$\text{mass adsorbed} = \text{mass adsorbed per unit rock mass} \times \text{total rock mass}$$

$$= K_d C \times \rho \times d \times 1 \times 1$$

$$= K_d C \rho d.$$

At the same time the amount in solution in that segment would be

$$\text{mass in solution} = \text{concentration} \times \text{volume}$$

$$= C \times \eta \times d \times 1 \times 1$$

$$= C \eta d.$$

Therefore, by the time the solute front was ready to "break through" into the second segment, the total solute mass in the first segment would be

$$\text{total mass} = K_d C \rho d + C \eta d.$$

How long would it take for this much solute mass to enter the first segment in the form of an input concentration C in water moving at a velocity v? The rate of entrance of such solute is just $C \eta v$ (since the cross-sectional area of each segment is unity), so the time required is

$$t_s = \frac{K_d C \rho d + C \eta d}{C \eta v} = \frac{K_d \rho d + \eta d}{\eta v}.$$

This is the time required for the solute to have filled up the first section and for the front to be ready to move into the next section. At the same time, the time required for the water itself to be ready to move into the next section is just the transit time over a distance d of water at a velocity v, namely,

$$t_w = \frac{d}{v}.$$

Therefore the ratio of the time required for the solute front to the time required by the water itself, which is the definition of the retardation factor R, is just

$$R = \frac{\frac{K_d \rho d + \eta d}{\eta v}}{\frac{d}{v}} = 1 + \frac{\rho}{\eta} K_d.$$

This is the desired formula. While we began this discussion by talking about a single section of length d, it should now be clear that the above logic just repeats itself for any number of such sections, so that the solute front is always moving at a rate R times slower than the water itself.

Note. The above argument could have been applied to an arbitrary length L instead of a "short" section d. We used the latter for several reasons: because of the tie-in to the figure in the test; because the assumption of instant equilibrium is more believable over a short segment; and because if one wants to make the argument more rigorous and less heuristic, then the limiting case of short segments will be needed.

4. Give a direct physical interpretation of the distribution coefficient K_d in terms of its typical units of milliliters per gram.

It can be interpreted as the number of milliliters of solution that would hold the same amount of solute as 1 gram of rock, under conditions of equilibrium. For example, if K_d were 50, say, for a particular solute and aquifer formation, this would say that under conditions of equilibrium, each gram of aquifer formation would hold the amount of solute present in 50 milliliters of ground water.

To verify this interpretation in terms of this example, use the definition of K_d as follows:

$$K_d = \frac{\text{mass of solute adsorbed on solid phase per unit mass of substrate}}{\text{concentration of solute in solution}}$$

$$50 = \frac{\text{mass of solute adsorbed on 1 gram of rock}}{\text{mass of solute dissolved in 1 milliliter of solution}}$$

$50 \times$ mass of solute dissolved in 1 milliliter of solution $=$ mass of solute adsorbed on 1 gram of rock

mass of solute dissolved in 50 milliliters of solution $=$ mass of solute adsorbed on 1 gram of rock

and from this it can be seen that the same logic applies to any K_d.

5.   Find the values of several retardation coefficients for specified solute/substrate combinations. (Hint: look in the radioactive waste literature, much of which is available on-line.)

A typical example obtained on-line is shown in Figure 5-D, and it illustrates several points raised in the text. For example, the K_d values range over about one and a half orders of magnitude, depending on the pH conditions of the ground water. These particular values are for cesium, which is an important component of nuclear waste and which has also been found to be leaking from various storage facilities connected with the atomic weapons program. Thus its movement through various soil and rock media has been heavily studied. The half life for the most important isotope, Cs-137, is 30 years, so that retardation in the ground-water system could in some cases delay it long enough for it to be reduced to negligible levels. The values shown on the graph would convert to fairly large R values, as can be seen from the conversion formula in the text or in the above exercises, although the precise values of density and porosity are not available.

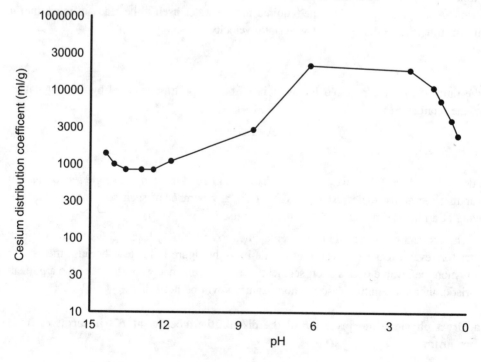

FIGURE 5-D

Example of cesium distribution coefficients for a certain geologic medium, as a function of pH

5.4 Determining the Hydraulic Head Contour Lines and General Flow Directions Using Data Available From Moderate or Large Numbers of Wells

This section is intended to be very "open-ended" and exploratory, something that we mathematics teachers may not generally provide much opportunity for in our usual course framework. It is not the intention here to lead the students to an optimal interpolation or contouring scheme, but rather to stimulate them to think precisely (e.g., making sure that their proposed algorithms are really well-defined) and creatively (e.g., not relying solely on one family of interpolating methods). They should be encouraged to formulate the problems here in a way that could be addressed by more detailed analysis, but they should also be encouraged to venture opinions and to discuss the advantages and disadvantages of various proposed approaches. Sometimes I take each group's proposed best method, and through several rounds of debate and voting (eliminating the less popular methods), I let them finally choose their favorite. Sometimes I go around the room and ask them to explain their vote, or I encourage teams to find a "fatal flaw" in another team's proposed approach.

If some of the students feel frustrated in the end because they've never been given the official answer about what is the best approach, perhaps this is all for the better. Maybe they will investigate the topic further or check to see how some commercial applications packages handle it.

A good theme to keep in mind while discussing these methods is the utility of looking at a simpler problem (e.g., one-dimensional interpolation) when trying to decide how to attack a more complex one.

1. Make a list of at least five distinct methods that a reasonable person might try to use to estimate the hydraulic head values at an arbitrary point in the geographic region shown in Figure 5-13. Be very specific in defining your methods so that they are completely "well-defined," meaning that a competent person should be able to carry them out for any given point in the region without asking you for further details about how the proposed method is supposed to work. (Such well-defined procedures are called *algorithms*.)

This list below combines the responses to Exercises 1 and 2. The methods are simply meant to be representative of different elementary approaches, and there is no special order. These examples illustrate the kinds of difficulties that can arise in this seemingly simple problem. Even though most groups of students are unlikely to come up with spline functions, they might be good to mention as they are in very common use, and they are easily discussed in a one-dimensional framework.

Method code: M1

Title:	Linear interpolation from three nearest points
Algorithm:	If the point is one of the data points, choose its value. Otherwise, choose the three nearest neighbors to the given point, breaking ties, if any, by choosing them in order of increasing values of the non-negative angle (with the positive horizontal direction formed by a ray from the given point to the potential grid points. If the three points are collinear, continue trying to pick a third point according to these criteria until obtaining one that is not collinear with the first two. Fit a plane to these three points (using the head values as the z-coordinate) and assign to the given grid point the corresponding z value from this plane.
Advantages:	Uses nearby points, which are most likely to be relevant to the unknown point's head value.
Disadvantages:	Gives a head surface that is not necessarily continuous and is not generally smooth. Unknown point may not be within the interior of the triangle formed by the three selected points, although it may be in the interior of triangles formed by other slightly more distant points.

Method code: M2

Title:	Linear interpolation from three surrounding points
Algorithm:	Begin by triangulating the region using the data points as nodes. (There is not a unique way to do this, but one can define an algorithm that will do it. For example, create a set of line segments by connecting every point to every other point, and then, working in decreasing order of length, eliminate segments from this set if they cross any other segment at a point interior to

either segment.) Now over each triangle construct the plane that interpolates to the z values at the vertices. This plane will be the interpolant over the given triangular portion of the region. There are no conflicts at vertices and boundaries of triangles (i.e., values are the same no matter which triangle you attach the point to). Values outside the convex hull of the data points (i.e., the set of points covered by such triangles) still need to be defined. For each of the four corners of the region, first define values there. Do that by choosing the nearest data point or segment to the corner (perpendicular distance), the latter if there are two equidistant, breaking ties as above by looking at the slope, then the triangle attached to that segment, or, if a point was chosen, the triangle attached to the data point which has the next nearest point to the corner; if there are two such, then choose the one with the nearest third point. Use the plane associated with that triangle to extrapolate a value for the corner. Once each corner has been determined, triangulate the new region (i.e., the area outside the original triangulated region) and apply the same method of linear interpolation to each of its triangular parts.

Advantages:	The resulting polyhedral surface is continuous. Surface trends are captured linearly.
Disadvantages:	The surface is not smooth at boundaries of triangles, and the corresponding contour lines would thus be polygonal lines, not smooth curves.

Method code: M3

Title:	Weighted average of nearby points
Algorithm:	If the point is one of the data points, simply use its given value. Otherwise pick the three nearest points. If there is a tie (such as if there is a fourth point at the same distance and one of these), take the extra point(s) also. Now calculate the z value at the given point as the weighted average of the z values of the selected grid points, using the reciprocal of the distances to calculate the weighting factors. That is, if n grid points (x_i, y_i) are selected for the evaluation of z at a given grid point (x, y), then letting

$$r_i = \frac{1}{d_i} = \frac{1}{\text{distance between } (x,y) \text{ and } (x_i, y_i)}$$

each weighting factor would be given by

$$w_i = \frac{r_i}{\sum_1^n r_i}$$

so that the weighted average would be

$$z = \sum_1^n w_i z_i.$$

Advantages:	Based on nearby points, which are likely to be the most relevant.
Disadvantages:	Not necessarily continuous. Does not capture surface trends. For example, if a point is not inside the polygon formed by the selected points, the calculated weighted average will still be within the range of their z values.

Method code: M4

Title:	Weighted average of all points
Algorithm:	Use the same weighting scheme as in method M3, but use every single data point in the evaluation of each new point.
Advantages:	Resulting surface is continuous.
Disadvantages:	Uses distant data points that are minimally relevant. Does not maintain trends well, especially outside convex hull of data points. Surface may not be smooth at data points.

Method code: M5

Title:	Linear regression
Algorithm:	Construct the planar surface that minimizes the sum of the squares of the residuals at all data points.
Advantages:	Continuous, smooth surface (a plane).

Disadvantages: Does not interpolate to the data points. (If the data are believed to have random measurement errors, this is not necessarily a disadvantage.) Contour lines will be parallel straight lines, probably an oversimplification.

Method code: M6

Title: Quadratic regression

Algorithm: Construct the quadratic surface

$$z = Ax^2 + Bxy + Cy^2 + Ex + Fy + G$$

with coefficients chosen to minimize the sum of the squares of the residuals.

Advantages: Continuous, smooth, curved surface. Contour lines can be determined analytically as conic sections, if desired.

Disadvantages: Does not interpolate to given data values. Restricted applicability to head surfaces that can be approximated reasonably well by quadratics.

Method code: M7

Title: Polynomial interpolation

Algorithm: Construct a two-dimensional polynomial of the form

$$z = \sum_{i,j} a_{i,j} x^i y^j$$

where i and j range from 0 to some numbers m and n, respectively. Choose m and n large enough so that there are enough coefficients to allow exact interpolation to all the data points. (For example, if there were 24 data points, substituting them into the above equation would yield 24 simultaneous linear equations for these coefficients. If m and n were each 3, there would only be 16 unknown coefficients $a_{i,j}$, and the system would likely be overconstrained and inconsistent.)

Advantages: Continuous, smooth, and interpolates exactly to each data point.

Disadvantages: Higher-order polynomial interpolants are notorious for having excessive fluctuations and oscillations, severely limiting their utility. (This can be seen with one-dimensional examples.) In addition, the form of the above equation would have to be changed in actual applications to use a different basis set of polynomials, or otherwise the linear system would generally be very ill-conditioned and prone to round-off error.

Method code: M8

Title: Two-dimensional spline

Algorithm: [Only an overview intended to give a flavor for spline interpolation.] Form a function of the form

$$z = \sum_{i,j} a_{i,j} p_i(x) q_j(y)$$

where the functions $p_i(x)$ and $q_j(y)$ are chosen from a given class of spline functions (e.g., piecewise cubic polynomials with continuity of the function and the first derivative at the nodal points, or under certain circumstances piecewise rational functions with specified continuity and smoothness properties); and the coefficients $a_{i,j}$ are chosen so that the above function interpolates to the given data at every point. Since the constraints translate into linear relationships among the coefficients, as long as there are enough such coefficients to satisfy the constraints, the problem reduces to solving a simultaneous set of linear equations.

Advantages: Smooth function that interpolates to all data points. Family of spline functions can be chosen to control other aspects (e.g., surface or contour line curvature or variability). Contour lines are smooth. Widely used in practice. (For further information, consult references on "NURBS," Non-Uniform Rational B-Splines.)

Disadvantages: Complex to apply, so much so that even some commercial programs have errors or undesirable properties. ("NURBS" has also been interpreted as "Nobody Understands Rational B-Splines"!)

2.   For each of the methods that you listed in Exercise 1, comment very briefly on any limitations or weaknesses that you think it might have in trying to estimate reasonably the value of the hydraulic head throughout the region shown. After completing these comments, rank order the proposed methods based on your own judgment about which ones you think are likely to give the most accurate results.

See answer to previous exercise.

3.   For your highest ranking method from Exercise 2, carry out the associated calculations to estimate the hydraulic head values at the following three points: A (3000, 4000); B (3500, 2000); C (1000, 500).

The results of applying the first six methods above are given in the following table. The seventh method as stated leads to a very ill-conditioned system; it has conceptual value only in the given form, and refinement is not warranted here based on the negative aspects mentioned. The eighth method would require extensive development that is beyond the present scope.

Method	$h(A)$	$h(B)$	$h(C)$	Notes
M1	237.60	224.60	206.06	
M2	237.73	224.60	205.39	Lower left corner $(0,0)$ is first determined to have value 199.27, and then $h(C)$ is determined using the triangle including the corner and the points $(500, 1000)$ and $(2500, 500)$
M3	248.15	224.60	208.39	The calculation of $h(A)$ is really an extrapolation, expected to be poor since weighted averages do not propagate trends. Four points are used as per the algorithm in the calculation of $h(B)$ and $h(C)$.
M4	231.64	225.67	219.23	
M5	237.03	229.48	194.86	
M6	238.24	224.85	210.38	

4.   Repeat the previous problem for your second-highest-ranked method, and see how the results compare with those using your highest-ranked method.

See the solution to the previous exercise.

5.   Based on your initial computational experience above, think once again, as creatively as possible, about what would be a good definition or criterion for a desirable method for estimating hydraulic head values in the present situation. See if you can formulate such a criterion in mathematical or quasi-mathematical terms. Based on your thoughts in connection with this problem, feel free to revise your list of methods originally proposed under Exercise 1. Furthermore, if you have ever studied any statistics, see if there are any lessons from that subject that you may want to take into consideration as you refine your list of possible methods here. (If you have not studied statistics, just ignore this suggestion.)

The statistics hint is intended to suggest regression methods as quite a distinct approach from actual interpolation at each data point. Such methods have been listed earlier. In general you want a continuous model for the head function, preferably a smooth one, and certainly one that does not violate any convincing physical expectations. A useful additional criterion takes the form of the principle of Occam's razor: you don't want your model to introduce any complications (e.g., any extra "twists" or strange behavior) not justified by the data. Unnecessary curvature would be a prime example of something to avoid. Some modelers (and others) know this as the KISS principle: "Keep It Simple, Stupid."

6.   Suppose that you now have an acceptable algorithm for generating estimated values of the head at any point within the region of interest (not just grid points). Design a conceptual algorithm that could then be used to generate a set of contour lines. Assume that the algorithm ultimately has to specify pairs of points between which straight line segments are to be drawn.

The objective here is not to be concerned about efficiency, but rather in just beginning with a logical approach that is well-defined and promising. (Keep in mind that for almost any algorithm, one can usually design pathological functions that will lead to poor results.)

Here is a basic sketch of a typical elementary approach: Assume that the region to be analyzed is rectangular. Define a very fine rectangular grid. Each four adjacent grid points form a small rectangle which we shall call a grid rectangle. Our contour lines will consist of sets of straight-line segments connecting points on the borders of the grid rectangles. A key tool will be the intermediate value theorem from calculus, applied to the unknown head value along sides of individual grid rectangles. In particular, if the value at one end of such a side is less than a, say, and at the other end it is greater than a, then at some point in between, it must equal a. Furthermore, we will choose as our approximate point where it equals a, the point obtained by linear interpolation between the two endpoints.

First determine the range of function values over the whole grid (a finite set of values to look at) and divide the range into a convenient number of values for which contour lines are to be drawn. For each such chosen value, say v, construct the contour line(s) for v as follows.

Number the grid rectangles and identify the subset of them that have v as a corner value (i.e., at a grid point) or where the intermediate value property implies a v value along one or more sides. Pick an arbitrary starting rectangle from this subset and connect any points with the value v. Cross this rectangle off the original list to indicate that it has been dealt with. Now extend a similar treatment to each of its neighbors in a methodical fashion, through common sides or corners that have a v value. For example, pick one side of the starting rectangle with a v value, and look at the adjacent rectangle, which shares a side. Connect the v point on that side to any other v point(s) in this second rectangle. Now cross this rectangle off the list. Repeat this process for all the other neighbors of the starting rectangle. (These could be called second-generation rectangles.) Now working with each of these, do the same for the third-generation rectangles, and so on, each time adding a rectangle only if it is still on the ("untreated") list. If at the end of this process, not all the rectangles have been crossed off the list, pick a new starting rectangle from the list and continue as before. (This would generally correspond to a distinct contour line with the same value.)

7. **Find a commercial computer program to which you have access that you can apply to the problem of estimating the head contour lines corresponding to the original data in Figure 5-13 and Table 5-1. [Hint: many quantitative programs can be applied or easily adapted to this problem, including math packages, statistical packages, and even some spreadsheet programs. This will generally be a two-step process: finding a way to approximate the head value at a general point, and then drawing a contour plot of this function. You might even want to use different programs for these two steps, depending on what you are most comfortable with.]**

A quadratic regression (using Mathcad) gives the estimated values shown in Table 5-A. If we compare those points for which we have actual data values, the corresponding residuals (data value minus calculated value) are found

TABLE 5-A
Estimated head values based on quadratic regression (Exercise 7)

5000									
4500	211.5	218.3	225.1	231.9	238.6	245.3	252.0	258.6	265.2
4000	207.2	213.5	219.7	225.9	232.1	238.2	244.3	250.4	256.4
3500	204.0	209.7	215.4	221.1	226.7	232.3	237.8	243.3	248.8
3000	201.9	207.1	212.2	217.3	222.4	227.4	232.4	237.3	242.3
2500	200.9	205.5	210.1	214.6	219.2	223.6	228.1	232.5	236.8
2000	201.0	205.1	209.1	213.1	217.1	221.0	224.9	228.7	232.5
1500	202.2	205.7	209.2	212.6	216.0	219.4	222.7	226.0	229.3
1000	204.5	207.5	210.4	213.3	216.2	219.0	221.7	224.5	227.2
500	208.0	210.4	212.7	215.1	217.4	219.6	221.8	224.0	226.2
0									
	0	500	1000	1500	2000	2500	3000	3500	4000	4500	5000

TABLE 5-B

Differences between data values and estimated values at data points (Exercise 7)

y \ x	0	500	1000	1500	2000	2500	3000	3500	4000	4500	5000
5000											
4500					-1.599			1.017		3.012	
4000		-0.878	-0.965					-0.547	-0.412		
3500		1.223								-1.799	
3000				1.102	-0.167					-2.263	
2500					0.955	0.244					
2000			1.928			0.55				-0.905	
1500					0.552				0.27		
1000		-1.349						0.466			
500						-1.662	-0.816			2.044	
0											

to have the values shown in Table 5-B. Corresponding head contour lines are shown in Figure 5-E superimposed on the original plot of the data.

8. **Sometimes the interpolation or extrapolation of given data can lead to incorrect results because of some key factor that is not represented by the limited set of data points provided. As an example of this situation, consider, within the context of Figure 5-13, the point** D **(3500, 3000). A cursory review of Figure 5-13**

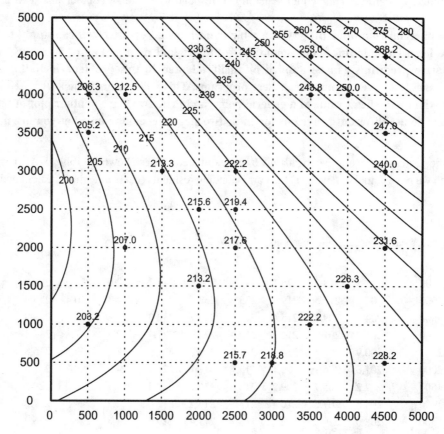

FIGURE 5-E

Head contour lines from quadratic regression to original head data (Exercise 7)

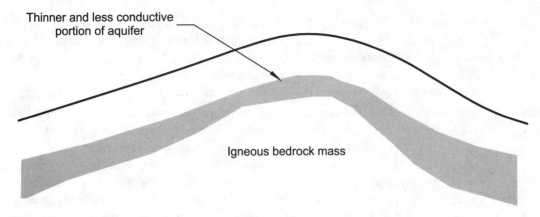

Thinner and less conductive portion of aquifer

Igneous bedrock mass

FIGURE 5-F
One possible situation leading to a ground-water high point

suggests that the unknown hydraulic head value at that point should be somewhere in the range between 222.2 and 240.0. Based on your knowledge of the relationship between geologic, topographic, and hydrologic values, provide at least one plausible physical situation that is both consistent with the data shown in Figure 5-13 but in which the actual hydraulic head value at point D would be greater than 250.

If the aquifer is a water-table aquifer, the water level could have a high point at this location due to the combination of underground deformation and perhaps restricted flow away from this point due to aquifer thinning or lower conductivity. (This would causing the kind of "backing up" we have discussed earlier.) Such a combination is suggested in Figure 5-F.

9. Suppose that the firm investigating the hydraulic head contours in the geographic region described by Figure 5-13 has sufficient funds available to drill one additional monitoring well. Identify on Figure 5-13 those areas where such a monitoring well would be most likely to reduce the amount of uncertainty associated with the contour lines. Explain your reasons. Could you define an objective quantitative criterion to capture your underlying logic? (If you completed Exercise 7, feel free to incorporate those results in your answer to this exercise.)

There is no absolutely correct answer to this exercise, but reasonable considerations would generally include those listed below. Note that sometimes such criteria are important beyond their scientific basis because they are needed to provide an objective form of justification for budget purposes or for making decisions in controversial cases. Interested readers might wish to investigate the topic of *geostatistics,* originally developed for ore reserve evaluations but now widely applied to hydrogeology. It provides techniques for separating out random errors from systematic trends and estimating the value of additional data.

Factor	Comments
Distance to data points	There would be more uncertainty about values that are farther from points where measurements have been taken. A quantitative criterion would be distance to the nearest data point.
Contour line curvature	Since flow lines are generally perpendicular to contour lines, areas in which there is substantial curvature in the latter are areas where flow lines would similarly be deformed. Since ground-water flow direction is perhaps the single most important output of interest, it is vital to pin down the flow directions and determine whether significant variations shown on a graph are due to real conditions or to the algorithm used to construct the graph. A quantitative criterion might be to choose the contour line with the maximum curvature and then to pick the point along this line such that the sum of its distances to the nearest three data points is maximal. In Figure 5-E, this would give a point on the left side of the lower boundary.

Calculated
residuals

Locations where the estimation procedure shows a relatively large residual when compared to the original data are likely to be places where the data is harder to fit, perhaps because of anomalous values. An additional data point in such regions can help one decide whether to modify the model. A quantitative criterion might be to put a new well at an untested grid point as close as possible to the grid point with the highest calculated residual.

Extrapolated vs.
interpolated
values

Estimated values for locations outside the "envelope" or convex hull of the original data points are likely to be more uncertain, as the mathematical constraints on estimates inside this region usually have the counterbalancing effect of causing them to diverge widely outside. (In the case of the weighted average methods, this is not true, as has been seen, but this method fails to carry trends outside the region of the data points at all, so the problem also holds with these methods.) Therefore, other factors being equal, it is generally more desirable to narrow down the boundary values than the interior values; and, of course, those on the downstream part of the flow would usually be more relevant in contamination cases. A quantitative criterion might be boundary points on the out-flowing part of the estimated flow regime that are most distant from any data points.

5.5 Exploration of the Relation Between Hydraulic Head Contour Lines and Ground-water Flow Lines

1.   **(Review of key concepts from basic physics.) Recall that a vector is an object that has magnitude and direction. Complete the following:**

a)   **Give an example of several quantities in physics that can be represented by vectors.**

   Force, velocity, acceleration, momentum.

b)   **Given two vectors, illustrate by means of a diagram how you can combine them to find the "resultant vector." (Hint: maybe you are familiar with this under the name "parallelogram rule for combining vectors.")**

   Both the addition and subtraction methods are illustrated in the left part of Figure 5-G. The "resultant" generally applies to the sum. The direction in which $A - B$ points is often confusing but can be resolved by looking at it as $A + (-B)$ and using parallelogram addition.

c)   **Given a vector drawn in the xy-plane, illustrate by means of an example how you can reverse the above process and represent the vector as the sum of two individual vectors, one pointing in the x-direction and one pointing in the y-direction.**

   This is illustrated on the right side of Figure 5-G.

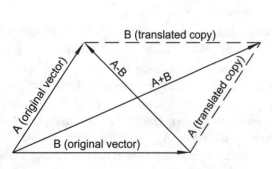 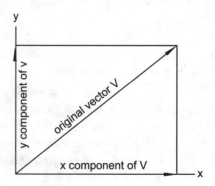

FIGURE 5-G
Geometric methods for combining and decomposing vectors

d) **In the previous part of this problem, you were asked to break a vector down into components in two given perpendicular directions, namely the x-direction and the y-direction. Is it possible, in general, to break a vector in the xy-plane down into components that are in two given directions, even if those two directions are not perpendicular to each other? If so, illustrate by means of a non-trivial example.**

We know it is possible because two non-parallel vectors in the plane are linearly independent and hence form a basis for the two-dimensional plane. From a geometric standpoint, look at it as the reverse of the addition process illustrated on the left side of Figure 5-G. We start with the long diagonal vector and want to break it down into vectors along the directions of A and B. To get the component in the A-direction (which in this special diagram naturally turns out to be A itself), draw a line parallel to B and through the tip of the original vector. Similarly for the other direction.

e) **If your answer to the previous part was "no," describe under what limited conditions you may actually be able to break down a vector into components in two non-perpendicular (but non-identical) directions. If your answer to the previous part was "yes," see if you can find a good reason why it may be preferable to break a vector down into perpendicular components rather than simply into general components in other directions.**

The underlying idea behind decomposing a vector into components is to do so in a way in which those components behave as independently as possible. For example, if you break a force vector down into two perpendicular components, then the force or acceleration in those two directions can be decoupled and analyzed separately. If the component directions were not perpendicular, such as A and B in Figure 5-G, the acceleration in the A-direction would not depend only on that component of the force vector because the B vector would also have a component in the A direction.

f) **Draw the following three vectors in the xy-plane: Vector 1, from the point (1, –3) to the point (5, –2); Vector 2, from the point (1, 1) to the point (3, 3); Vector 3, from the point (–1, 2) to the point (–4, 5). Can you break down Vector 1 into two components along the two directions represented by Vectors 2 and 3? If so, do it, being sure to explain exactly what your logic is for each step. If not, explain why this is impossible. Be sure to work with good diagrams.**

See Figure 5-H. This question is intended to lead into the next one involving the geometric representation of vectors as arrows in the plane. In particular, the set of all arrows in the plane, wherever located, but with the same direction and the same length, form an equivalence class. Addition and subtraction can be performed with any convenient members of each class, and the results are independent of the members chosen. Therefore in this figure, equivalent vectors have been positioned with their starting points all at the origin (although that particular

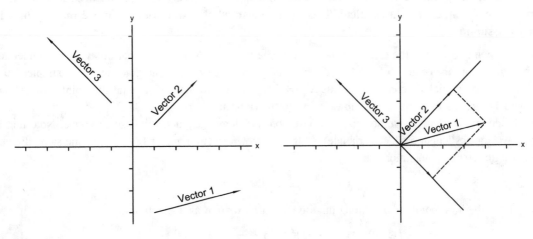

FIGURE 5-H
Repositioning of vectors to illustrate decomposition

common point is not even necessary), and the decomposition process can be carried through as described earlier. Since Vectors 2 and 3 happen to be perpendicular, this simply involves dropping a perpendicular from the tip of Vector 1 to each of the lines along which these other two vectors lie.

g) **As in the previous part of this problem, arrows in the xy-plane can be used to represent two-dimensional vectors. Consider the following two vectors: Vector 1, from the point (1, 1) to the point (2, 4); and Vector 2, from the point (4, −1) to the point (5, 2). Do these two arrows represent different vectors or exactly the same vector? Be very precise in your explanation.**

As discussed in the previous part, two arrows with same magnitude and direction, as here, are simply different representatives of the same equivalence class of arrows and represent precisely the same vector.

2. **(Review of selected multivariable calculus topics.) Answer the following questions in the context of a general function of two variables of the form $z = f(x, y)$ that is assumed to be "smooth" (meaning that it has continuous first partial derivatives on its domain):**

a) **What is meant by the "directional derivative" of such a function? Describe your answer in words, not equations, but be precise.**

At any given point, it is the rate of change of the function value with respect to distance in some specified direction. Note that if you have smooth function and you were to consider a direction 180 degrees opposite, then the directional derivative in the new direction would be the negative of the original value.

b) **What is meant by the "gradient" of such a function?**

The gradient is a two-dimensional vector whose two components are the partial derivatives of f with respect to x and y respectively.

c) **Is the directional derivative at a point a scalar value or a vector?**

A scalar.

d) **Is the gradient at a point a scalar value or a vector?**

A vector.

e) **Give a typical example of a non-trivial function of the type being discussed in this problem. (You will use this function for later parts of this problem.)**

$$f(x, y) = x^2 + xy$$

f) **Give a typical example of the calculation of the directional derivative of this function, and also an example of the calculation of the gradient of this function, both of them at a given point. If you need any additional assumptions or input values for these calculations, make clear what assumptions you are making or what values you are assuming.**

Consider the point $(3, 5)$ and the direction specified by the vector $(-3, 4)$. (Note that this vector is understood to refer to the arrow from the origin to $(-3, 4)$ and hence its equivalence class.) The gradient ∇f is simpler, so we do it first. It is seen by partial differentiation to be $\nabla f = (2x + y, x)$, so that its value at the point in question is just $(11, 3)$, which happens to point steeply upward and to the right in the xy-plane.

For the directional derivative, we do this here based on the elementary definition given above. Note that the given direction vector has length 5 (3, 4, 5-triangle), so that the new direction vector $(\frac{-3}{5}, \frac{4}{5})$ has the same direction but is of unit length. Hence points along the ray

$$(3, 5) + s(\tfrac{-3}{5}, \tfrac{4}{5}), \quad s \geq 0$$

are distance s from the base point $(3, 5)$. Along this ray f can be written as a function of s

$$f(x, y) = (3 + \tfrac{-3}{5}s)^2 + (3 + \tfrac{-3}{5}s)(5 + \tfrac{4}{5}s)$$

$$= 24 - \tfrac{21}{5}s - \tfrac{3}{25}s^2$$

so that its derivative at 0, corresponding to the point in question, can be found by elementary differentiation to be $\frac{-21}{5}$. Note that even if $f(s)$ were not differentiable at 0, but had only a right-hand derivative at $s = 0$, this would still fall within the meaning of directional derivative. Note also that if we repeated this calculation in the opposite direction, we would indeed get the negative of the above answer.

g) Use dot product notation to provide a formula for the directional derivative of a function at a point in terms of the gradient at that point and a vector pointing in the direction in which you wish to calculate the directional derivative.

If P is the point at which we are working and V is the given direction for the directional derivative, the latter has the value

$$\nabla f(P) \cdot \frac{V}{\|V\|}.$$

The division by the length or norm of V is needed to convert the direction to a unit vector. Applying this formula to the previous example, we have:

$$(11, 3) \cdot (-\tfrac{3}{5}, \tfrac{4}{5}) = -\tfrac{21}{5}.$$

h) What special property does the directional derivative at a point have when taken in the specific direction represented by the gradient? (Hint: you may wish to use the representation of the dot product in terms of the cosine of some angle in order to help you answer this question.)

The gradient direction gives a maximum value for the directional derivative. To see this, use the cosine representation of the dot product in the definition of the directional derivative:

$$\nabla f(P) \cdot \frac{V}{\|V\|} = \|\nabla f(P)\| \times \left\|\frac{V}{\|V\|}\right\| \times \cos\theta$$

which makes it obvious that the maximum occurs when $\theta = 0$. (And its value is the norm of the gradient, since the middle factor is unity.)

i) Consider a point along an arbitrary level line (= contour line) of the function f. Prove that the gradient is perpendicular to the tangent to that level line at the given point. Use your previously defined function to present an actual numerical example to verify this as well.

(The question has real meaning only if the gradient is not the 0 vector.) Along a level line, the function is constant. Hence the directional derivative is 0 in either direction along the level line (or, strictly speaking, in a direction along the tangent to the level line at any point). From the cosine formula for the directional derivative, this can only happen if $\theta = 90°$ or $270°$, in which case the gradient direction would be perpendicular to the tangent direction, as required.

j) The statement is occasionally made that "the gradient points in the direction of steepest ascent." Explain this statement. Furthermore, give a corresponding characterization of the direction of steepest descent.

The directional derivative has been seen above to be a maximum in the gradient direction; this is what "steepest ascent" refers to. The negative or reverse of the gradient direction would be the direction of steepest descent. This point of view will be used later.

3. (Review of key linear algebra concepts for readers with some linear algebra background.) Refresh your memory on the situation addressed in parts c, d, and e of Exercise 1, above. Address these same issues in the language of linear algebra, especially the concepts of spanning set or basis.

Some of this has already been mentioned above, so let us look at it here in n dimensions. In an n-dimensional vector space, any n linearly independent vectors must also span the space and hence form a basis. When these basis elements are orthogonal (= perpendicular) it is easy to use the dot product to find the individual coefficients for the representation of a given vector as a linear combination of the basis elements. For example, in this case, if a vector V is desired to be written as a linear combination of basis vectors B_i in the form

$$V = c_1 B_1 + c_2 B_2 + c_3 B_3 + \cdots + c_n B_n,$$

we can find each coefficient c_i by taking the dot product of V and B_i:

$$
\begin{aligned}
V \cdot B_i &= (c_1 B_1 + c_2 B_2 + c_3 B_3 + \cdots + c_n B_n) \cdot B_i \\
&= c_1 B_1 \cdot B_i + c_2 B_2 \cdot B_i + c_3 B_3 \cdot B_i + \ldots + c_n B_n \cdot B_i \\
&= c_i B_i \cdot B_i \quad \text{(since all other terms are 0 by orthogonality)} \\
&= c_i (B_i \cdot B_i)
\end{aligned}
$$

so that

$$c_i = \frac{V \cdot B_i}{B_i \cdot B_i}.$$

If the basis set is orthonormal (orthogonal plus norm of each element equal to unity), then even the denominator drops out in this last expression.

4.   Imagine that you are standing on the side of a mountain, with the only peculiar aspect of this mountain being that its surface is quite smooth rather than being covered with vegetation or rocks. In mathematical terms, you could of course think of the mountain as the surface representing the graph of the function f described at the beginning of these exercises. Now imagine that you take a hockey puck and lay it down at your feet on the side of the mountain so that it can begin to slide downhill. In what precise direction will the puck start to move? (Hint: don't just say "down" because there are many different directions it could go that would be generally downward. And don't even just say "in the steepest downward direction" unless you can provide a precise argument why this would be the case.) Give a convincing explanation of your answer. You may base it either on the physics-type reasoning represented in Exercise 1, or you may use some of that reasoning in combination with the terminology and concepts reviewed in Exercise 2.

This is a more difficult problem than it may seem. One has to work hard to be precise and to avoid large jumps in logic.

Think of the puck as being located at a point P in 3-space located on the side of the mountain. The mountain, being a smooth surface, can be described by a set of level lines in the domain space, which we can think of as a horizontal xy-plane at or under the base of the mountain. For a given level line, the set of corresponding points on the surface form a curve, all of whose z values are the same. Call this the corresponding level curve in 3-space.

There are two forces operating on the hockey puck, the downward force of gravity and the reactive force exerted on the puck by the mountain. The gravitational force W points straight down, so it is perpendicular to any horizontal line through P. Therefore it is perpendicular to the tangent to the level curve in 3-space passing through P. Equivalently, the gravity force vector drawn at P would lie within the plane through P that is perpendicular to this level-curve in 3-space through P. This plane is drawn in Figure 5-I. The level curve through P would come right out of the paper perpendicular to the figure.

Since this force lies within this plane, it has no components that would exert any force on an object to move other than within the given plane. The force W can now be broken down into two components, one perpendicular to the surface and one tangent to it, but both lying within the plane of the figure, as shown. The force of the mountain on the puck is only reactive to these. In particular, the mountain certainly exerts a force that exactly counterbalances the perpendicular component (or else the puck would sink right through the rock!) If friction is present, the mountain may also exert a force in a direction exactly opposite to the tangential component of W. As long as the puck slides, this means that the tangential force of the mountain is less than the tangential component of W. In any case, the net force on the puck is as shown in the figure, and so the direction of movement will be the same.

Since the puck begins to move from P in the plane represented by the figure, and since the plane projects onto a line in the xy-plane we imagined at the base of the mountain, the puck's movement, if tracked in this xy-plane,

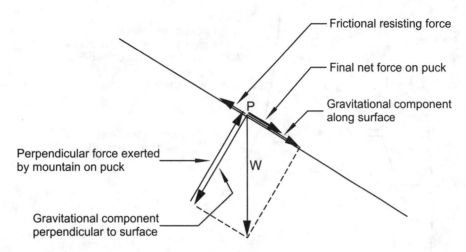

FIGURE 5-1
Resolution of forces on hockey puck within plane perpendicular to level curve in 3-space through P

will initially move only along this line. But this line of motion is perpendicular to the level line in the xy-plane because that is directly under the level curve in 3-space that passes through P.

Therefore, in the xy-plane, the line of motion is perpendicular to the level line. But we also know that the gradient vector is perpendicular to the level line. Therefore the line of motion and the gradient vector must be collinear.

5. Give a convincing argument why it is that in an isotropic geologic medium the direction of ground-water flow at any point would be perpendicular to the hydraulic head contour line at that point. Be careful not to make any assumptions in your analysis without providing justification. Furthermore, be sure to understand where you are making use of the assumption that the geologic medium is isotropic.

A two-dimensional treatment of this question is probably most consistent with our discussion of ground-water flow, and it is easy to follow. Therefore assume that the aquifer flow is essentially horizontal and described by movement on an xy-plane. Darcy's law essentially says that flow is driven by a change in head; if there is no change, there will be no flow. (This is just as in a simple electrical circuit: if there is no voltage, there will be no current.)

If you pick a point P in the plane describing the aquifer and look at the head contour line through that point, the directional derivative in a direction tangent to the contour line at P will be 0. Hence there is no force in that direction to bring about flow. Therefore, if we resolve the net force causing flow of a particle of water at P into components along the contour line and perpendicular to the contour line, the second component will contain the entirety of the force. Furthermore, if the medium is isotropic, it will not resist this force with a force in any preferential direction, but rather one that is only reactive to the given force but in the opposite direction. (This is similar to the force exerted by the mountain on the puck in the previous example.) Hence the net force will be in the same direction, as will the movement of the water. Therefore the movement will be perpendicular to the contour line at every point. (This direction must then also be collinear with the gradient direction.)

6. In this exercise you are asked to sketch in a conceptual way the general flow lines you would expect to encounter in an anisotropic situation.

a)   Begin by drawing a large square filling out most of a page, and within it draw a typical set of hydraulic head contour lines. (You have seen many examples of these in previous sections.)

b)   Go along each of the individual contour lines and sketch in a small arrow pointing in the direction of "steepest descent," which, as you know (if you have worked the multivariable calculus exercises in this

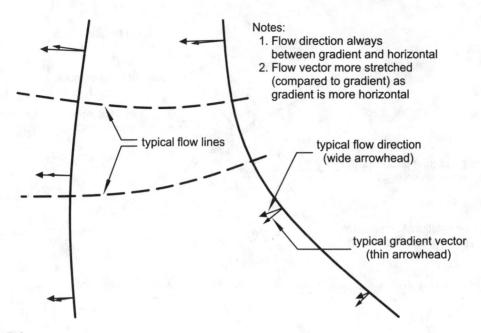

FIGURE 5-J
Flow directions in an anisotropic situation

section), is the negative of the direction of the gradient vector. In drawing in these arrows, make the length of the arrow roughly proportional to the magnitude of the gradient vector. (Hint: if you have locations where the contour lines are closer together, meaning that the surface is somewhat steeper, the gradient is larger so the arrows should be somewhat longer.)

c)   Assume that in the two-dimensional region shown by your diagram ground water can move roughly twice as easily in the x-direction as in the y-direction. Taking this fact into account, with a different-color writing instrument from the one with which you drew the previous set of arrows, draw an additional arrow at the base of each of the previous arrows to represent a vector of estimated fluid flux at that point. Do your best to take into account both the magnitude of the driving force and the anisotropic nature of the medium in drawing these new vectors.

d)   Using the fluid flux vectors as a guide, sketch in a set of estimated flow lines for this situation.

   Figure 5-J illustrates this situation and indicates the key characteristics resulting from the anisotropy.

5.6 The Continuous Version of Darcy's Law in Two and Three Dimensions For Isotropic and Anisotropic Media

1.   Suppose that the head function in the vicinity of a well that is withdrawing water is given by the function

$$h(x, y) = 8 \ln \left(10 + \sqrt{x^2 + y^2} \right).$$

Assume that the hydraulic conductivity of the aquifer is 30 ft/day, and consider the three points $(10, 15)$, $(-20, 30)$, and $(5, -5)$. (Hint: Begin by drawing the contour lines and the expected flow lines. Minimal or no computation should be necessary to do this.)

   The contour lines are just concentric circles with center at the origin. The flow lines are the radii of the circles, with flow inward toward the center where the head is the least.

a)   Calculate the hydraulic gradient vector at each of these points.

The partial derivatives are given by

$$h_x = \frac{8}{10 + \sqrt{x^2 + y^2}} \times \frac{1}{2\sqrt{x^2 + y^2}} \times 2x$$

$$h_y = \frac{8}{10 + \sqrt{x^2 + y^2}} \times \frac{1}{2\sqrt{x^2 + y^2}} \times 2y$$

so that the gradient vectors turn out to be:

$$\nabla h(10, 15) = (.16, .24)$$

$$\nabla h(-20, 30) = (-.10, .14)$$

$$\nabla h(5, -5) = (.33, -.33).$$

b)   Calculate the flux vector at each of these points and specify the units for each of its components.

The units for each component will be the same as those for K, namely ft/day, since the partial derivatives are dimensionless (ft/ft). Therefore the values are:

$$q(10, 15) = (4.75, 7.12)$$

$$q(-20, 30) = (-2.89, 4.34)$$

$$q(5, -5) = (9.94, -9.94)$$

c)   Calculate the directional derivative of the head at each of the points in the direction of flow.

These values are found in the last column of the spreadsheet table below:

x	y	sqrt sum	h	hx	hy	u1	u2	dir der
10	15	18.028	26.666	0.158	0.237	-0.555	-0.832	-0.285
-20	30	36.056	30.639	-0.096	0.145	0.555	-0.832	-0.174
5	-5	7.071	22.699	0.331	-0.331	-0.707	0.707	-0.469

(The u's in this table are the components of the unit vectors in the direction from the point to the origin, which are just the normalized negatives of the (x, y) vectors in this case.) The directional derivative should indeed be negative in the direction of flow since the head must be decreasing in that direction.

Now evaluate whether each of the following statements is true or false, giving a logical rationale for your answer, and use your calculated values at all three points to illustrate the situation:

d)   The magnitude of the flux vector is proportional to the magnitude of the gradient vector.

True, since we would expect the flux to equal the magnitude of the directional derivative times the hydraulic conductivity K. This is because flow is locally one-dimensional along an axis represented by a flow line. The following spreadsheet table verifies this:

x	y	hx	hy	q1	q2	mag grad	mag flux	ratio
10	15	0.158	0.237	4.750	7.125	0.285	8.563	30
-20	30	-0.096	0.145	-2.891	4.336	0.174	5.211	30
5	-5	0.331	-0.331	9.941	-9.941	0.469	14.059	30

e)   The magnitude of the flux vector is proportional to the directional derivative in the gradient direction.

This is true for the same reason as above, noting that the magnitude of the gradient vector is the same as the magnitude of the directional derivative.

Give a precise physical interpretation of the following:

f)   The magnitude of the flux vector.

It has units of ft/day, which actually should be interpreted as ft³/day per square foot of cross-sectional area perpendicular to the flow path. That is, it gives the flow rate through a unit cross section perpendicular to the axis of flow.

g)   Each individual component of the flux vector. (Make sure that there is clear meaning to what you are saying. Use a diagram to illustrate if necessary.)

These give the flow rates through unit cross sections that are not perpendicular to the flow direction, but oriented perpendicular to the x- and y-axes respectively. Such a unit cross section could be drawn in space, and then we could collect all the water that crossed it in a given amount of time, even if it crossed it at an oblique angle. That would give us the rate.

2.   The quantities represented by q and by ∇h in the above are referred to as *vector fields*. In other terminology, they might be called functions from R^2 to R^2, or vector-valued functions of two variables. The key concept is that for any input value (x, y), each of these expressions has an output value which also contains two components. The arrow representation used in Figure 5-16 is a useful vehicle for drawing such vector fields. The question you are going to be asked here is somewhat subtle, and you should attempt it only if you have a good grounding in multivariable calculus and mathematical derivations. In developing Darcy's law for the isotropic case and for two dimensions as above, we assumed that the vector q was proportional to the gradient vector ∇h at every point. Suppose we were to replace that assumption with the possibly weaker assumption that the magnitude of the vector q at any point (x,y) is proportional to the magnitude of the gradient vector. If we combine this weakened assumption with the other assumption made above, namely that q and ∇h always point along the same line (and, of course, in the opposite direction), does it still follow that q must have the form of a constant K times the negative of the gradient vector?

Yes. To see this, note that the fact that flow is always in the negative gradient direction can be written mathematically as:

$$q(x, y) = -M(x, y)\nabla h(x, y)$$

where $M(x, y)$ is a scalar function of x and y that we know is always positive. The proportionality condition in the problem says that there is some constant K such that

$$\|q(x, y)\| = K\|\nabla h(x, y)\|.$$

K could not be negative here except in the trivial case when everything is 0 (and no water is going anywhere). Using the first equation to replace q in the second, we have:

$$\|-M(x, y)\nabla h(x, y)\| = K\|\nabla h(x, y)\|$$

$$|-M(x, y)|\,\|\nabla h(x, y)\| = K\|\nabla h(x, y)\|$$

$$M(x, y) = K$$

which completes the argument.

3.   Make your best guess as to how Darcy's law would need to be reformulated for the anisotropic situation just described.

The natural answer to this, which would be a very reasonable guess, would be the one discussed in the text right after the exercise. As is further discussed in the text, however, it is not generally true, although it does apply to a very wide range of practical anisotropic problems where the directions of anisotropy are orthogonal (especially vertical vs. horizontal).

4.   Verify that this more general anisotropic form of Darcy's law does indeed satisfy the two requisite conditions listed just previously. (Hint: depending on the extent of your background in matrix theory, you may want to work with either the matrix or scalar form of the equations.)

Any linear transformation of the form $u = Av$, where u and v are vectors and A is a matrix, satisfies these conditions. The first condition is just that $A \cdot 0 = 0$, where 0 is the 0 vector. The second is just that $A \cdot kv = k \cdot Av = ku$, where k is a scalar (positive in our case), combined with the fact that $\|ku\| = k\|u\|$ and similarly for v.

5. **[Recommended only for readers with a good background in mathematical proofs.] Is the linear transformation given above the *only* functional form that satisfies the requisite two conditions listed?**

No, condition 2 (which subsumes condition 1) does not imply that we have a linear transformation. In equivalent terms, suppose we want to investigate possible functions F that map points $X = (x, y)$ to points $S = (s, t)$ and that satisfy condition 2. In fact, without loss of generality we can even assume that the constant of proportionality is unity, so that in this case we are looking for the range of functions that map the plane to the plane and preserve the norms of vectors. Therefore, every circle centered at the origin would be mapped into itself. While a rotation through a fixed angle would be a linear transformation (and hence have the matrix representation shown), if we scale the angle of rotation by, say, the norm of the vector, we can easily construct a nonlinear transformation that still satisfies the given conditions. For example, let S equal a rotation of X by an angle equal to the norm of X. Then in fact,

$$S = \|X\| \left(\cos \|X\|, \sin \|X\| \right).$$

(The second factor here, the vector, has norm unity since it is on the unit circle.) This is not a linear transformation, as can be seen from looking at the images of two simple points:

$$(0, 1) \rightarrow (\cos 1, \sin 1) \approx (.54, .54)$$
$$2(0, 1) = (0, 2) \rightarrow 2(\cos 2, \sin 2) \approx (-.83, 1.82) \neq 2\,(.54, .54)$$

which is just the nonlinearity of the trig functions. Since it is not linear, it does not take the form of matrix multiplication (since the latter is easily seen to satisfy this linearity condition).

6. **Identify at which of the two boundaries of the region R_2 in Figure 5-17 fluid flow through the boundary can actually occur, and through which two boundaries fluid flow cannot occur. Explain your reasoning.**

It cannot flow through the dashed-line boundaries, which are themselves flow lines, since flow lines cannot cross. It can cross the solid-line boundaries, as these are contour lines.

7. **[Requires knowledge of eigenvalues and eigenvectors of matrices.] The hydraulic conductivity matrix K discussed above is generally "symmetric," meaning that $K_{12} = K_{21}$. (The reasons for this are too complex for discussion here.) Assuming that this is the case, describe a coordinate transformation such that if the entire flow problem is reformulated in terms of these new coordinates u and v, the off-diagonal elements of the new but corresponding hydraulic conductivity matrix are actually 0. Explain why, in this new coordinate system, one might comment that "the flow in the u direction and the flow in the v direction are essentially decoupled." (These new directions for u and v, if indicated on the diagram of the original problem in x and y, would be referred to as the "principal directions" for the problem.) Describe the nature of the new coordinate system. For example, are u and v perpendicular in the xy-plane?**

The following is true even for n dimensions and is standard linear algebra material. For a (real) symmetric matrix, the eigenvalues (i.e., the roots of the characteristic polynomial) are all real numbers (instead of possibly having imaginary components), not necessarily distinct; and there exists a complete basis for n-space with each basis element being an eigenvector and the set of such eigenvectors being an orthogonal set (i.e., they are pairwise orthogonal). By scaling, each such eigenvector can be chosen to have unit length. If the coordinates for n-space are changed to this new orthonormal basis, the new matrix representation of the linear transformation represented by the original matrix will be a diagonal matrix whose diagonal entries are the eigenvalues.

This corresponds to the original suggested form of Darcy's law for two dimensions, discussed in the text just after Exercise 3. It corresponds to identifying those "principal directions" in which the flow can indeed be treated one dimension at a time by simple analogs of the one-dimensional Darcy equation.

8. With reference to the previous problem, now consider a situation in which the original x- and y-directions are already the principal directions, so that the hydraulic conductivity matrix in Darcy's law for this particular situation already has 0's as the off-diagonal entries. Find an expression for the angle between the flow direction and the negative gradient direction in terms of the directional hydraulic conductivities and/or the hydraulic head function. Simplify as much as possible so as to identify the factors that affect this angle.

The angle between the flow lines and the negative gradient direction is simply the angle between the vectors (q_1, q_2) and $(-h_x, h_y)$, which can be determined using our two different formulas for this dot product:

$$(q_1, q_2) \cdot (-h_x, -h_y) = (-K_1 h_x, -K_2 h_y) \cdot (-h_x, -h_y)$$

$$K_1 h_x^2 + K_2 h_y^2 = \left(\sqrt{(K_1 h_x)^2 + (K_2 h_y)^2} \right) \left(\sqrt{h_x^2 + h_y^2} \right) \cos\theta$$

$$\theta = \cos^{-1} \left(\frac{K_1 h_x^2 + K_2 h_y^2}{\left(\sqrt{(K_1 h_x)^2 + (K_2 h_y)^2} \right) \left(\sqrt{h_x^2 + h_y^2} \right)} \right)$$

$$= \cos^{-1} \left(\frac{\beta^2 + \alpha}{\sqrt{(\beta^2 + \alpha^2)(\beta^2 + 1)}} \right)$$

where $\alpha \equiv K_2/K_1$, $\beta \equiv -h_x/h_y$, and we assume for this last equation that $h_y \neq 0$. The value of α is constant throughout, but the value of β, which is the slope of the contour line at any point where it is not vertical, will, in general, vary from point to point. These are precisely the ratios one would expect to enter into the determination of deviation due to anisotropy. (Note in the above that if the system is actually isotropic, so that $\alpha = 1$, the angle comes out correctly as $\theta = 0$.)

9. In the same context as the previous problem, consider an arbitrary unit direction vector $u = (u_1, u_2)$. Show that a one-dimensional generalized Darcy-type law applies to the flow rate, with the general form

$$q_u = -K_u \frac{dh}{ds},$$

where q_u in this equation is to be interpreted as the fluid flux in this direction and K_u is an equivalent one-dimensional hydraulic conductivity (which may not be constant) for that direction. Find an expression for this hydraulic conductivity K_u in terms of the directional hydraulic conductivities and/or the hydraulic head function, and discuss when it would be constant throughout a region.

The ordinary flux vector q will point along a flow line, and since the u direction is an arbitrary direction, we need to find the component of q is this direction. This is simply the magnitude of the projection of q onto u, which is just their dot product in this case. Furthermore the expression $\frac{dh}{ds}$ in the problem is just the directional derivative in the direction of u, which can also be represented by a dot product. Putting these ideas in equation form, we have:

$$q_u = q \cdot u = -K_1 h_x u_1 - K_2 h_y u_2$$

$$\frac{dh}{ds} = \nabla h \cdot u = h_x u_1 + h_y u_2$$

so that K_u would have to have the form below:

$$K_u = \frac{-K_1 h_x u_1 - K_2 h_y u_2}{-h_x u_1 - h_y u_2} = \frac{K_1 h_x u_1 + K_2 h_y u_2}{h_x u_1 + h_y u_2}$$

$$= K_1 \left(\frac{\beta - \alpha \tan \theta}{\beta - \tan \theta} \right)$$

where θ is the angle of the unit vector u, and, as in the solution to the previous problem, $\alpha \equiv K_2/K_1$, $\beta \equiv -h_x/h_y$, and we assume for this last equation that $h_y \neq 0$. (As mentioned earlier, the value of α is constant throughout, but the value of β, which is the slope of the contour line at any point where it is not vertical, will, in general, vary from point to point.) Because of the presence of β, it is clear that the so-called "equivalent hydraulic conductivity" is constant, in general, only in regions where the slope of the contour lines is constant, that is, when the contour lines form a set of parallel lines. (Then the flow lines also form a set of parallel lines at some fixed angle to the contour lines.)

10. State the three-dimensional version of Darcy's law for both the isotropic and anisotropic cases.

For the isotropic case, it would have the form $q = -K \nabla h$, where K is a scalar constant, q is a 3-dimensional vector function of (x, y, z), and h is a scalar function of (x, y, z). For the anisotropic case, it would still have the same general form, the only difference being that K would now be a 3×3 hydraulic conductivity matrix allowing for cross-effects among all three directions.

11. Suppose that the head function in the vicinity of a well (located at the origin), which is withdrawing water, is given by the function

$$h(x, y) = 55 + 12 \ln(10 + \sqrt{x^2 + 2y^2}).$$

Suppose further that the aquifer is isotropic with a hydraulic conductivity value of 15 ft/day and a porosity of 0.25.

a) Sketch the head contour lines and describe them geometrically.

The head contour lines are the lines along which the quantity $x^2 + 2y^2$ is constant, which form a family of ellipses centered at the origin. These and some further aspects of this problem are shown in Figure 5-K.

b) Find the equation of the flow line along which water would travel from the point $(100, 150)$ to the well.

The flow line will everywhere be perpendicular to the contour lines and hence to the ellipses in the above family. Therefore, except along the axes, where either the flow lines or the contour lines would be vertical, we get the slope of the contour lines first by implicit differentiation as follows:

$$x^2 + 2y^2 = C$$

$$2x + 4y \frac{dy}{dx} = 0$$

$$\frac{dy}{dx}(\text{contour line}) = \frac{-x}{2y},$$

and then we take the negative reciprocal to get the slope of the flow lines:

$$\frac{dy}{dx}(\text{flow line}) = \frac{2y}{x}.$$

In the first quadrant, where we are operating because of our initial point, this yields a general solution as follows:

$$\frac{dy}{y} = \frac{2dx}{x}$$

$$\ln y = 2 \ln x + D$$

$$y = e^{2 \ln x + D}$$

$$y = Ax^2.$$

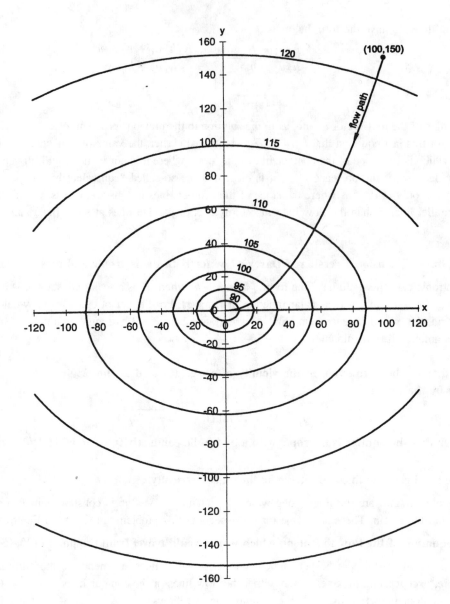

FIGURE 5-K
Head contour lines and flow path for Exercise 11

The initial condition $y(100) = 150$ yields a value for the constant $A = 0.015$, so we may write our solution as

$$y = .015x^2.$$

c) Find the travel time required by the ground water to move along this flow line from the point (100, 150) to the well.

We will set this up as an integral over the x interval. The time to traverse the portion of the curve corresponding to an x increment of Δx (say from $x + \Delta x$ to x) is given by

$$t(x) = \frac{\text{distance}}{\text{velocity}} = \frac{\sqrt{(\Delta x)^2 + (\Delta y)^2}}{-\dfrac{K}{\eta}\dfrac{dh}{ds}}$$

where the directional derivative in the denominator is in the direction of flow and is determined from the original head function using the usual dot product:

$$\frac{dh}{ds} = \nabla h \cdot \left(\frac{-\Delta x}{\sqrt{(\Delta x)^2 + (\Delta y)^2}}, \frac{-\Delta y}{\sqrt{(\Delta x)^2 + (\Delta y)^2}} \right)$$

$$= h_x \frac{-\Delta x}{\sqrt{(\Delta x)^2 + (\Delta y)^2}} + h_y \frac{-\Delta y}{\sqrt{(\Delta x)^2 + (\Delta y)^2}}.$$

This enables the simplification of the expression for the incremental time:

$$t(x) = \frac{\sqrt{(\Delta x)^2 + (\Delta y)^2}}{-\dfrac{K}{\eta} \left(h_x \dfrac{-\Delta x}{\sqrt{(\Delta x)^2 + (\Delta y)^2}} + h_y \dfrac{-\Delta y}{\sqrt{(\Delta x)^2 + (\Delta y)^2}} \right)}$$

$$= \frac{(\Delta x)^2 + (\Delta y)^2}{\dfrac{K}{\eta} h_x \Delta x + \dfrac{K}{\eta} h_y \Delta y}$$

$$= \frac{1 + \left(\dfrac{\Delta y}{\Delta x} \right)^2}{\dfrac{K}{\eta} h_x + \dfrac{K}{\eta} h_y \dfrac{\Delta y}{\Delta x}} \Delta x.$$

Thus the total travel time would be the integral of this expression from 0 to 100, which we can evaluate numerically using any standard calculus program. In particular,

$$T = \int_0^{100} \frac{1 + \left(\dfrac{dy}{dx} \right)^2}{\dfrac{K}{\eta} h_x + \dfrac{K}{\eta} h_y \dfrac{dy}{dx}} \, dx = \frac{1}{60} \int_0^{100} \frac{1 + (.03x)^2}{h_x + h_y(.03x)} \, dx$$

where, along the given trajectory, h_x and h_y are given by

$$h_x = \frac{12}{10 + \sqrt{x^2 + 2(.015x^2)^2}} \times \frac{x}{\sqrt{x^2 + 2(.015x^2)^2}}$$

$$h_y = \frac{12}{10 + \sqrt{x^2 + 2(.015x^2)^2}} \times \frac{2(.015x^2)}{\sqrt{x^2 + 2(.015x^2)^2}}.$$

The value of the integral is found to be 24.7 days.

5.7 Laplace's Equation and Inverse Problems

1. **Rewrite the two- and three-dimensional versions of Laplace's equation using the common del (∂) notation for partial derivatives.**

$$\frac{\partial^2 h}{\partial x^2} + \frac{\partial^2 h}{\partial y^2} = 0; \qquad \frac{\partial^2 h}{\partial x^2} + \frac{\partial^2 h}{\partial y^2} + \frac{\partial^2 h}{\partial z^2} = 0.$$

2. **Beginning with the above mass balance equation, derive the two-dimensional Laplace equation. (Hint: if you are not sure how to proceed, review the derivation for the one-dimensional situation in Section 5.2.)**

We divide the equation in the text by the product $\Delta x \Delta y$, apply Darcy's law in each direction, divide by K, and then take the limit as both increments approach 0:

$$0 \approx \frac{q_1(x - \Delta x, y) - q_1(x + \Delta x, y)}{2\Delta x} + \frac{q_2(x, y - \Delta y) - q_2(x, y + \Delta y)}{2\Delta y}$$

$$0 \approx \frac{-Kh_x(x - \Delta x, y) + Kh_x(x + \Delta x, y)}{2\Delta x} + \frac{-Kh_y(x, y - \Delta y) + Kh_y(x, y + \Delta y)}{2\Delta y}$$

$$0 = \lim_{\substack{\Delta x \to 0 \\ \Delta y \to 0}} \left[\frac{-h_x(x - \Delta x, y) + h_x(x + \Delta x, y)}{2\Delta x} + \frac{-h_y(x, y - \Delta y) + h_y(x, y + \Delta y)}{2\Delta y} \right]$$

$$0 = h_{xx}(x, y) + h_{yy}(x, y).$$

We have used the central difference formulas for the derivative in taking these limits, rather than the ones more commonly encountered in the definition of the derivative in elementary calculus. These approximations are discussed in further detail in Section 5.8.

3.   Find three distinct, non-trivial examples of functions that satisfy the Laplace equation in all or some portion of the xy-plane.

By trial and error one can find examples, such as the following (where a and b are arbitrary constants):

$$ax + by + c, \quad axy, \quad x^3 - 3xy^2, \quad \frac{x}{x^2 + y^2}, \quad a(x^2 - y^2) + bxy, \quad \sin x \cosh x.$$

In addition, it is easy to see that the partial derivatives of harmonic functions are harmonic (assuming the higher order derivatives exist).

4.   For one of the three functions you identified in the previous exercise, pick a relatively simple region of the plane in which it is defined, such as a square, rectangle, or circle, and determine the actual boundary conditions that it exhibits on the boundary of this region. (The more common situation would be to begin with the boundary conditions and try to find the unknown harmonic function satisfying those conditions, but this exercise simply asks you to look at the situation in reverse in order to see some typical boundary conditions.)

The first example clearly has linear values on the boundary of a rectangle, say, and they are continuous at the corners since the function itself is continuous. Next look at the function xy on the boundary of the square region defined by the points $x = \pm 1$, $y = \pm 1$. It is also linear along the borders, equaling, for example, the function $h(x, y) = x$ along the top border where $y = 1$. This same function along the unit circle takes the values $\cos\theta \sin\theta = \frac{1}{2}\sin 2\theta$ so it completes 2 full cycles in one revolution around the border. The function $\frac{x}{x^2 + y^2}$, considered on the unit circle, reduces to the value of x, so it takes the values $\cos\theta$ as you go around the circle. For smaller circles, it takes on bigger and bigger extremes, especially on the x-axis, and it is not even defined at the origin.

5.   The derivation of the two-dimensional Laplace equation was based on the assumption of constant hydraulic conductivity throughout the region of interest. Assume instead in this exercise that the hydraulic conductivity gradually increases to the right within the region shown in Figure 5-18. In particular, assume that the hydraulic conductivity function can be written in the following form:

$$K(x, y) = K_0 + \alpha x.$$

Derive the partial differential equation that would need to be satisfied by the hydraulic head $h(x, y)$ in this case. (The aquifer is still assumed to be isotropic with respect to flow properties.)

Returning to the derivation completed within Exercise 2, the key changes are shown below:

$$0 \approx \frac{-(K_0 + \alpha(x - \Delta x))h_x(x - \Delta x, y) + (K_0 + \alpha(x + \Delta x))h_x(x + \Delta x, y)}{2\Delta x}$$

$$+ \frac{-(K_0 + \alpha x)h_y(x, y - \Delta y) + (K_0 + \alpha x)h_y(x, y + \Delta y)}{2\Delta y}$$

$$0 = \lim_{\substack{\Delta x \to 0 \\ \Delta y \to 0}} \left[(K_0 + \alpha x)\frac{-h_x(x - \Delta x, y) + h_x(x + \Delta x, y)}{2\Delta x} + \alpha \Delta x \frac{h_x(x - \Delta x, y)}{2\Delta x} \right.$$

$$\left. + \alpha \Delta x \frac{h_x(x + \Delta x, y)}{2\Delta x} + (K_0 + \alpha x)\frac{-h_y(x, y - \Delta y) + h_y(x, y + \Delta y)}{2\Delta y} \right]$$

$$0 = h_{xx}(x, y) + h_{yy}(x, y) + \frac{\alpha}{(K_0 + \alpha x)}h_x(x, y).$$

6. Return to the derivation of the two-dimensional Laplace equation, but assume that the aquifer is anisotropic with the x- and y-directions being principal directions with respective hydraulic conductivities K_1 and K_2 that remain constant throughout the region of interest. What partial differential equation would need to be satisfied by h in this case?

The equation is derived below, also following logic similar to that used above:

$$0 \approx \frac{-K_1 h_x(x - \Delta x, y) + K_1 h_x(x + \Delta x, y)}{2\Delta x} + \frac{-K_2 h_y(x, y - \Delta y) + K_2 h_y(x, y + \Delta y)}{2\Delta y}$$

$$0 = \lim_{\substack{\Delta x \to 0 \\ \Delta y \to 0}} \left[K_1 \frac{-h_x(x - \Delta x, y) + h_x(x + \Delta x, y)}{2\Delta x} + K_2 \frac{-h_y(x, y - \Delta y) + h_y(x, y + \Delta y)}{2\Delta y} \right]$$

$$0 = K_1 h_{xx}(x, y) + K_2 h_{yy}(x, y).$$

After all the complications seen in the previous section related to the flow field in the anisotropic case, this is actually a very encouraging result because it shows that the head distribution in the anisotropic case can probably be obtained almost as easily as in the isotropic case because the above equation can clearly be re-scaled so that it becomes Laplace's equation.

7. Draw an appropriate diagram and provide a complete derivation of the three-dimensional Laplace equation for the head function $h(x, y, z)$ in an isotropic aquifer with uniform hydraulic conductivity K and in which flow is three-dimensional.

Rather than draw a new diagram, we can actually use the same diagram used for the two-dimensional case (Figure 5-18 in the text), except with the one modification that in the third dimension, perpendicular to the figure, the box will now have thickness $2\Delta z$, rather than 1. There are now potential flows through all six faces, and the mass balance takes the form:

$$0 = \text{left face flow in} + \text{right face flow in}$$

$$+ \text{bottom face flow in} + \text{top face flow in}$$

$$+ \text{far face flow in} + \text{near face flow in}$$

$$0 \approx q_1(x - \Delta x, y, z)(2\Delta y \cdot 2\Delta z) - q_1(x + \Delta x, y, z)(2\Delta y \cdot 2\Delta z)$$

$$+ q_2(x, y - \Delta y, z)(2\Delta x \cdot 2\Delta z) - q_2(x, y + \Delta y, z)(2\Delta x \cdot 2\Delta z)$$

$$+ q_3(x, y, z - \Delta z)(2\Delta y \cdot 2\Delta z) - q_3(x, y, z + \Delta z)(2\Delta y \cdot 2\Delta z).$$

Now we apply Darcy's law to each dimension, divide through by $2\Delta x 2\Delta y 2\Delta z$, divide through by K, and then take the limit as all three increments go to 0:

$$0 \approx \frac{-Kh_x(x-\Delta x,y,z)+Kh_x(x+\Delta x,y,z)}{2\Delta x}$$

$$+\frac{-Kh_y(x,y-\Delta y,z)+Kh_y(x,y+\Delta y,z)}{2\Delta y}+\frac{-Kh_z(x,y,z-\Delta z)+Kh_z(x,y,z+\Delta z)}{2\Delta z}$$

$$0 = \lim_{\substack{\Delta x \to 0 \\ \Delta y \to 0 \\ \Delta z \to 0}}\left[\frac{-h_x(x-\Delta x,y)+h_x(x+\Delta x,y)}{2\Delta x}\right.$$

$$\left.+\frac{-h_y(x,y-\Delta y)+h_y(x,y+\Delta y)}{2\Delta y}+\frac{-h_z(x,y,z-\Delta z)+h_z(x,y,z+\Delta z)}{2\Delta z}\right]$$

$$0 = h_{xx}(x,y)+h_{yy}(x,y)+h_{zz}(x,y).$$

8.   Consider a two- or three-dimensional steady-state problem involving one of the sets of quantities in Table 5-2 other than those relating to fluid flow. For this situation, draw appropriate diagrams and provide a direct derivation of Laplace's equation. (Hint: the key issue here is to convert the flow equations and the conservation condition from the ground-water derivation to physically meaningful statements in terms of the new variables.)

Let us do this for Fick's law in three dimensions, since this relates very closely to material in the next chapter. Here the concentration $C(x,y,z)$ takes the place of hydraulic head, the diffusion coefficient D takes the place of hydraulic conductivity, and q is the flux of the dissolved or gaseous material. For a three-dimensional box fixed in space and of dimensions $2\Delta x \times 2\Delta y \times 2\Delta z$, the net rate of movement of the material of interest into the box is the sum of the rates through all six faces, namely:

left face flow in + right face flow in

+ bottom face flow in + top face flow in

+ far face flow in + near face flow in

$$\approx q_1(x-\Delta x,y,z)(2\Delta y \cdot 2\Delta z) - q_1(x+\Delta x,y,z)(2\Delta y \cdot 2\Delta z)$$

$$+ q_2(x,y-\Delta y,z)(2\Delta x \cdot 2\Delta z) - q_2(x,y+\Delta y,z)(2\Delta x \cdot 2\Delta z)$$

$$+ q_3(x,y,z-\Delta z)(2\Delta y \cdot 2\Delta z) - q_3(x,y,z+\Delta z)(2\Delta y \cdot 2\Delta z)$$

$$\approx -DC_x(x-\Delta x,y,z)(2\Delta y \cdot 2\Delta z) + DC_x(x+\Delta x,y,z)(2\Delta y \cdot 2\Delta z)$$

$$- DC_y(x,y-\Delta y,z)(2\Delta x \cdot 2\Delta z) + DC_y(x,y+\Delta y,z)(2\Delta x \cdot 2\Delta z)$$

$$- DC_z(x,y,z-\Delta z)(2\Delta y \cdot 2\Delta z) + DC_z(x,y,z+\Delta z)(2\Delta y \cdot 2\Delta z).$$

Note that, unlike the previous case, we did not set this equal to 0, although we will in a moment. That is because in the diffusion case it is less common for the system to reach a steady-state where the net flow into the box is 0, except after the system has reached some kind of equilibrium where the concentration is constant everywhere. However, it is indeed possible if material continues to be supplied or removed at some boundary points. But note that in the absence of steady state conditions, the given expression would be the time rate of change of the total inventory within the box, namely, $D_tC(x,y,z) \times 2\Delta x \times 2\Delta y \times 2\Delta z$ where we are now implicitly thinking of C also as a function of time t. Just to keep this derivation on the more general level described here, let us use this derivative as the left side of the mass balance. Now, dividing both sides by $2\Delta x 2\Delta y 2\Delta z$ and taking the limit as all increments go to 0, we obtain

$$C_t = DC_{xx} + DC_{yy} + DC_{zz}$$

which is the three-dimensional diffusion partial differential equation that arises in Chapter 6! In the case of some steady-state concentration distribution, so that the left side is 0, we divide through by the diffusion constant and obtain Laplace's equation.

9. Explain the basis for this last sentence.

The higher head at P indicates that material is "backing up" in the aquifer, in the sense discussed previously. That is, it is having a harder time getting through, just like when an ice-jam or log-jam blocks a river, and so the water builds up to a higher level or higher head until the increased gradient allows enough to get through to accommodate the flow from upstream and maintain a steady-state.

10. Let $h(x, y)$ and $g(x, y)$ represent solutions to Laplace's equation in the two subregions under discussion in connection with Figure 5-20, and assume that these functions do satisfy the relevant boundary conditions for hydraulic head along their respective portions of the exterior border of the overall square region. What additional consistency condition(s) must apply to h and g along the slanted border between the two subregions in order for these solutions to Laplace's equation to form a physically meaningful solution to the overall unknown hydraulic head distribution throughout the large region?

At each point of the boundary between the subregions, the head values and the flux vectors must be the same (since the second system must "pick up where the first leaves off") . In mathematical terms, let K and L represent the two respective hydraulic conductivities in the regions corresponding to h and g. Our conditions imply that for each boundary point between the regions:

$$h(x, y) = g(x, y)$$
$$Kh_x(x, y) = Lg_x(x, y)$$
$$Kh_y(x, y) = Lg_y(x, y).$$

11. Use the divergence concept to derive the Laplace equation for the case of steady-state ground-water flow.

For any point in the flow regime, the integral on the left side of the divergence theorem must be 0 over any spherical region surrounding the point. Taking the limit as the radius of such a region approaches 0, clearly the divergence value at the point itself must be 0, for if it were not, by continuity of the divergence at the center point, we could choose such a region around the point where it always had the same sign as at the center point, and the integral over that particular region would not be 0. Thus the divergence of the flux field q must be 0 everywhere, and hence so must the divergence of the head gradient field since it differs from the former only by a multiplicative constant. But this is Laplace's equation, since:

$$0 = \text{div}\,(\nabla h) = \text{div}(h_x, h_y, h_z) = h_{xx} + h_{yy} + h_{zz}.$$

5.8 Introduction to Numerical Modeling

1. Show that the central difference approximation to the derivative is simply the average of the forward and backward approximations.

We calculate the required average and see that it turns out to be the central difference formula:

$$\frac{\dfrac{f(x + \Delta x) - f(x)}{\Delta x} + \dfrac{f(x) - f(x - \Delta x)}{\Delta x}}{2} = \frac{f(x + \Delta x) - f(x - \Delta x)}{2\Delta x}.$$

2.   [For readers who are familiar with Taylor Series and Taylor's Theorem.] Use Taylor's Theorem with Remainder or Taylor Series to expand the expressions $f(x + \Delta x)$ and $f(x - \Delta x)$ around the value x itself; and then, by plugging these expressions into the formulas for the forward, backward, and central difference formulas, demonstrate that the central difference formula has a "higher order of convergence." (This means that as Δx approaches 0, this approximation would be expected to converge to the true derivative value much more rapidly.) Construct a numerical example for a specific function and use it to demonstrate this conclusion.

The first- and second-order Taylor expansions of $f(x + \Delta x)$ around the point x are given by

$$f(x + \Delta x) = f(x) + f'(x)\Delta x + \mathcal{O}(\Delta x)^2$$

$$f(x + \Delta x) = f(x) + f'(x)\Delta x + \frac{f''(x)}{2}(\Delta x)^2 + O(\Delta x)^3,$$

where the remainder is represented using the $\mathcal{O}$ notation, which is an abbreviation for some function bounded by a constant times the expression following the $\mathcal{O}$, this applying for Δx sufficiently small. Plugging these formulas into the forward and central difference formulas, respectively, we obtain:

$$\frac{f(x + \Delta x) - f(x)}{\Delta x} = f'(x) + \mathcal{O}(\Delta x)^1$$

$$\frac{f(x + \Delta x) - f(x - \Delta x)}{2\Delta x} = f'(x)\Delta x + \mathcal{O}(\Delta x)^2,$$

since the f'' terms cancel out in the second case. Note that the power of x has been reduced by one in each of the $\mathcal{O}$ terms because of division through by the denominator on the left. (The backward difference formula has the same behavior as the forward formula.) Note now that the forward and backward formulas would converge to the derivative value with an error of order Δx, whereas the central formula converges with error of order $(\Delta x)^2$, which is much faster.

As an example, consider the exponential function e^x at the point $x = 0$, where the derivative has the value 1. Let us take $\Delta x = .1$. The forward and central difference formulas yield the values:

$$\frac{f(x + \Delta x) - f(x)}{\Delta x} = \frac{e^{.1} - 1}{.1} = 1.05171$$

$$\frac{f(x + \Delta x) - f(x - \Delta x)}{2\Delta x} = \frac{e^{.1} - e^{-.1}}{.2} = 1.00167,$$

which confirms the superiority of the central difference formula.

3.   [Involves Taylor's Theorem or Taylor series.] Analyze the quality of the above approximation to $f''(x)$ by using Taylor expansions, as earlier, and determine what the lowest power of Δx is in an expression for the difference between $f''(x)$ and this approximation. (This power is called the *order* of the approximation, and the error being evaluated here is called the *truncation error*.)

We simply plug in the Taylor expansion for the terms in the approximation, keeping in mind that the terms represented by the O notation may change from place to place and that it is only the bounding property that is being represented by this notation:

$$\frac{f(x + \Delta x) - 2f(x) + f(x - \Delta x)}{(\Delta x)^2} = \frac{f'(x)\Delta x + \frac{f''(x)}{2}(\Delta x)^2 + \mathcal{O}(\Delta x)^3 - f'(x)\Delta x + \frac{f''(x)}{2}(\Delta x)^2 + \mathcal{O}(\Delta x)^3}{(\Delta x)^2}$$

$$= f''(x) + \mathcal{O}(\Delta x)^1,$$

from which we conclude that this is a first order approximation.

4.   Construct reasonable approximations to the partial derivatives h_x, h_y, **and** h_{xy} **at the point** (x, y)**. If for one or more of these approximations you feel you need to assume that the** h **values are also available at additional grid points analogous to those in Figure 5-22, you may state and use such an assumption. (Note**

that these partial derivatives do not show up in Laplace's equation; however, they do show up in many other important partial differential equations governing ground water and other environmental problems.)

Central difference formulas for the first derivatives are

$$h_x(x,y) \approx \frac{h(x+\Delta x, y) - h(x - \Delta x, y)}{2\Delta x}$$
$$h_y(x,y) \approx \frac{h(x, y + \Delta y) - h(x, y - \Delta y)}{2\Delta y}.$$

For the mixed derivative, let us use a central formula in the y-direction, applied to the function h_x, and approximate this latter function by the central formula above, although applied at the points with y values equal to $y \pm \Delta y$:

$$h_{xy}(x,y) \approx \frac{h_x(x, y + \Delta y) - h_x(x, y - \Delta y)}{2\Delta y}$$

$$\approx \frac{\dfrac{h(x+\Delta x, y+\Delta y) - h(x-\Delta x, y+\Delta y)}{2\Delta x} - \dfrac{h(x+\Delta x, y-\Delta y) - h(x-\Delta x, y-\Delta y)}{2\Delta x}}{2\Delta y}$$

$$\approx \frac{h(x+\Delta x, y+\Delta y) - h(x-\Delta x, y+\Delta y) - h(x+\Delta x, y-\Delta y) + h(x-\Delta x, y-\Delta y)}{4\Delta x \Delta y}.$$

If not all these corner points were available, meaning that we did not have function values at all of them, we could get by with fewer by using forward or backward difference formulas at either stage, but the resulting approximations would in general be poorer.

5. **Find the unknown head values at the grid points in Figure 5-23.**

If we write these 25 simultaneous equations in the usual matrix form $Ax = b$, then A has the values shown in Table 5-C, and the vector b, written in tabular form to conserve space, has the form

122.5	70	75	78.75	175
55	0	0	0	88.75
53.75	0	0	0	87.5
52.5	0	0	0	86.25
103.75	56.25	61.25	67.5	160

where the entries should be read sequentially column by column and match up with the layout of grid values as well. In particular, these are all simply one-fourth of the sum of any boundary values adjacent to the corresponding grid point. The nine 0's are for the nine interior grid points.

Now we obtain the solution vector by any standard routine. If we re-format the 25 values in the solution vector to match the original layout of the grid points, the values in the following 5×5 table result:

251.21	272.03	292.62	312.78	335.92
242.82	264.29	285.68	307.58	330.89
235.77	256.63	278.22	300.96	325.05
228.62	248.26	269.61	293.00	318.35
220.45	238.17	258.96	283.08	310.36

Note that these look very reasonable in light of the averaging principle mentioned earlier, as well as in terms of the expected head distribution for the boundary values shown in Figure 5-23.

6. **With respect to your solution to the previous problem (and even if you never got to the point of actually calculating numerical values), reiterate in simple physical terms what you would expect a typical h value, say h_{17}, to represent. Did you need to make any assumption about the specific value of hydraulic conductivity K to carry out the above solution? Summarize the assumptions of the model on which your calculated h value are based.**

TABLE 5-C
Coefficient matrix for Exercise 5

```
 1   -.25  0    0    0   -.25  0    0    0    0    0    0    0    0    0    0    0    0    0    0    0    0    0    0    0
-.25  1   -.25  0    0    0   -.25  0    0    0    0    0    0    0    0    0    0    0    0    0    0    0    0    0    0
 0   -.25  1   -.25  0    0    0   -.25  0    0    0    0    0    0    0    0    0    0    0    0    0    0    0    0    0
 0    0   -.25  1   -.25  0    0    0   -.25  0    0    0    0    0    0    0    0    0    0    0    0    0    0    0    0
 0    0    0   -.25  1    0    0    0    0   -.25  0    0    0    0    0    0    0    0    0    0    0    0    0    0    0
-.25  0    0    0    0    1   -.25  0    0    0   -.25  0    0    0    0    0    0    0    0    0    0    0    0    0    0
 0   -.25  0    0    0   -.25  1   -.25  0    0    0   -.25  0    0    0    0    0    0    0    0    0    0    0    0    0
 0    0   -.25  0    0    0   -.25  1   -.25  0    0    0   -.25  0    0    0    0    0    0    0    0    0    0    0    0
 0    0    0   -.25  0    0    0   -.25  1   -.25  0    0    0   -.25  0    0    0    0    0    0    0    0    0    0    0
 0    0    0    0   -.25  0    0    0   -.25  1    0    0    0    0   -.25  0    0    0    0    0    0    0    0    0    0
 0    0    0    0    0   -.25  0    0    0    0    1   -.25  0    0    0   -.25  0    0    0    0    0    0    0    0    0
 0    0    0    0    0    0   -.25  0    0    0   -.25  1   -.25  0    0    0   -.25  0    0    0    0    0    0    0    0
 0    0    0    0    0    0    0   -.25  0    0    0   -.25  1   -.25  0    0    0   -.25  0    0    0    0    0    0    0
 0    0    0    0    0    0    0    0   -.25  0    0    0   -.25  1   -.25  0    0    0   -.25  0    0    0    0    0    0
 0    0    0    0    0    0    0    0    0   -.25  0    0    0   -.25  1    0    0    0    0   -.25  0    0    0    0    0
 0    0    0    0    0    0    0    0    0    0   -.25  0    0    0    0    1   -.25  0    0    0   -.25  0    0    0    0
 0    0    0    0    0    0    0    0    0    0    0   -.25  0    0    0   -.25  1   -.25  0    0    0   -.25  0    0    0
 0    0    0    0    0    0    0    0    0    0    0    0   -.25  0    0    0   -.25  1   -.25  0    0    0   -.25  0    0
 0    0    0    0    0    0    0    0    0    0    0    0    0   -.25  0    0    0   -.25  1   -.25  0    0    0   -.25  0
 0    0    0    0    0    0    0    0    0    0    0    0    0    0   -.25  0    0    0   -.25  1    0    0    0    0   -.25
 0    0    0    0    0    0    0    0    0    0    0    0    0    0    0   -.25  0    0    0    0    1   -.25  0    0    0
 0    0    0    0    0    0    0    0    0    0    0    0    0    0    0    0   -.25  0    0    0   -.25  1   -.25  0    0
 0    0    0    0    0    0    0    0    0    0    0    0    0    0    0    0    0   -.25  0    0    0   -.25  1   -.25  0
 0    0    0    0    0    0    0    0    0    0    0    0    0    0    0    0    0    0   -.25  0    0    0   -.25  1   -.25
 0    0    0    0    0    0    0    0    0    0    0    0    0    0    0    0    0    0    0   -.25  0    0    0   -.25  1
```

A typical calculated head would be an estimate of the unknown head value in the aquifer at the location corresponding to the grid point. The specific value of K does not enter into the numerical solution, as expected, since it does not enter Laplace's equation. (It was divided out in the derivation.) The underlying model assumptions are that the geologic medium is isotropic with respect to conductivity and that K is constant throughout the region.

7. **Use a spreadsheet program or other computer program to carry out the following iterative method for solving the same problem as in Exercise 5. Make an initial guess at all 25 unknown values in the grid. (It does not have to be a good guess. In fact, feel free to try this problem with a crazy guess. You will still get the right answer, but it may take a little longer.) Your first iteration involves replacing each value with the average of its four nearest neighbors. Your next iteration repeats that process on the new values. Continue this process over and over until the values seem to be "converging" (that is, there is little change from step to step). (Hint: Try to automate this process with your computer so that you can perform a large number of iterations conveniently.) How do your answers compare with those from Exercise 5?**

We shall start with the values (including the boundary values) shown below:

240	260	280	300	315	340	365
230	300	300	300	300	300	360
220	300	300	300	300	300	355
215	300	300	300	300	300	350
210	300	300	300	300	300	345
205	300	300	300	300	300	340
200	210	225	245	270	300	335

although we could even start with numbers way out of the range, such as 0's for all the unknown points. After ten iterations of the type described, we have

240.00	260.00	280.00	300.00	315.00	340.00	365.00
230.00	253.51	275.64	296.29	315.50	337.28	360.00
220.00	246.87	270.80	292.24	312.47	333.40	355.00
215.00	240.69	264.39	286.19	306.94	328.11	350.00
210.00	232.98	255.33	276.79	298.46	321.18	345.00
205.00	223.07	242.32	263.28	286.34	312.05	340.00
200.00	210.00	225.00	245.00	270.00	300.00	335.00

where you can observe poorer convergence near the center, which is more removed from the effect of the boundary conditions. After 50 iterations, we have

240.00	260.00	280.00	300.00	315.00	340.00	365.00
230.00	251.22	272.04	292.63	312.79	335.92	360.00
220.00	242.83	264.31	285.70	307.59	330.90	355.00
215.00	235.78	256.66	278.25	300.98	325.06	350.00
210.00	228.63	248.28	269.63	293.02	318.36	345.00
205.00	220.45	238.18	258.98	283.09	310.36	340.00
200.00	210.00	225.00	245.00	270.00	300.00	335.00

and this shows excellent agreement with the results of the previous method.

8. Most iterative methods involve rewriting an equation as a "fixed point problem," $x = g(x)$, and then generating successive approximations to the solution by the formula $x_{n+1} = g(x_n)$, starting from some initial guess x_0.

a) When Newton's method from elementary calculus is interpreted in this way, what is the corresponding function g?

Newton's method is a method for seeking solutions to the equation $f(x) = 0$, but this original equation is rewritten in the strange fixed point form

$$x = x - \frac{f(x)}{f'(x)}$$

corresponding to the function

$$g(x) \equiv x - \frac{f(x)}{f'(x)}.$$

b) When g is an ordinary (scalar) function, not necessarily the one from Newton's method, show that if the derivative of g at the unknown solution is close to 0, the method will converge faster, at least if you start close enough to that solution.

If x is an actual solution and the x_n's are a sequence of successive iterates as described in the problem, then the distances between each iterate and the actual solution satisfy:

$$|x_{n+1} - x| = |g(x_{n+1}) - g(x)| = |g'(c_n)(x_n - x)| \le M|x_n - x| \le M^n|x_1 - x|$$

where we have applied the mean value theorem and have assumed that the constant M bounds the derivative of g on the interval where all the x values lie. If M is less than 1, then the iterates certainly converge to that solution. If M is much less than 1, the convergence is quite fast since the factor M^n approaches 0 fast.

c) For Newton's method, what is the value of the derivative of g at the unknown solution? (This should give an idea of why Newton's method converges so fast.)

The value is 0 at that particular point, since

$$g'(x) = 1 - \frac{\left(f'(x)\right)^2 - f(x)f''(x)}{\left(f'(x)\right)^2} = 1 - \frac{\left(f'(x)\right)^2 - 0 \cdot f''(x)}{\left(f'(x)\right)^2} = 1 - 1 = 0.$$

Hence (assuming continuity of the derivative), for some neighborhood of the point the value can be kept arbitrarily small. This explains why Newton's method, once it gets close to a solution (which it does not always do), really zooms in fast.

d)   [For readers with a strong linear algebra background.] The same iterative concepts can be applied to matrix equations of the form $Ax = b$, where A is a square matrix and x and b are vectors. This is just the case of n equations in n unknowns. If you were to find several ways to rewrite such a system in the fixed point form $x = Dx + c$, so as to try iterations or successive approximations, what qualities do you think D should have so as to encourage rapid convergence?

In general terms, where G is not necessarily matrix multiplication, the relationship analogous to the earlier case is

$$\|x_{n+1} - x\| = \|G(x_{n+1}) - G(x)\| \le M\|x_n - x\| \le M^n \|x_1 - x\|$$

where M would be a bound on the norm of the Jacobian matrix $G_x = \left[\dfrac{\partial G_i}{\partial x_j}\right]$, and the matrix norm chosen could be any matrix norm consistent with the vector norm being used. For the case as stated in the problem, it is slightly easier to look at it as follows:

$$\|x_{n+1} - x\| = \|Dx_{n+1} - Dx\| = \|D^{n+1}x_1 - D^{n+1}x\| = \|D^{n+1}(x_1 - x)\| \le \|D^{n+1}\|\|x_1 - x\|,$$

so the question really is what would make the higher and higher powers of D converge to the 0 matrix. A theorem in linear algebra proves that such convergence is equivalent to the spectral norm of D being strictly less than one, where the *spectral norm* $\rho(D)$ is defined to be the maximum of the complex moduli of all the (possibly complex) eigenvalues of D. This is certainly to be reasonably expected since we think of the eigenvalues as the factors by which certain components of an input vector are multiplied during transformation by D. So the final conclusion is that we would like to find a D with small eigenvalues, certain all strictly within the unit circle.

9.   Show that the method used in Exercise 7 can be interpreted in the framework of Exercise 8d. In particular determine the corresponding matrix D. Also find an alternative iterative scheme, using a different matrix D, although you need not carry out the actual iterative calculations to test its convergence.

Certainly the most obvious approach would be to modify the system as follows:

$$Ax = b$$
$$x = (I - A)x + b \equiv G(x).$$

Here the matrix D is just $D = I - A$, where A is the matrix whose values were given earlier in Table 5-C. If you generate successive iterates by transforming your original initial guess in Exercise 7 according to the above function, that is, $x_{n+1} = (I - A)x_n + b$, you will simply be repeating the process you carried out in that earlier exercise as can be verified by tracking the matrix multiplication and addition. For example, with respect to the value of h_1 at any step, this depends on the first row of $I - A$, which has all 0's except for values of $1/4$ in the second and sixth columns. Thus the matrix term contributes the values $\dfrac{h_2}{4} + \dfrac{h_6}{4}$ to the next iterate. The b term contributes the remaining expected values, namely $\dfrac{230}{4} + \dfrac{260}{4}$. This method is called the *point Jacobi method*.

One alternative group of iteration schemes would derive from writing A as the sum of two other matrices, $A = E + F$, with at least one of them, say E, being chosen to be non-singular (= invertible). Then we could rewrite the system as follows:

$$Ax = b$$

$$(E + F)x = b$$

$$Ex = -Fx + b$$

$$x = -E^{-1}Fx + E^{-1}b \equiv G(x).$$

Our previous case corresponded to $E = I$, the identity matrix. Such decompositions into two or more terms have been extensively studied with a view towards reducing the spectral radius of the effective matrix D and hence speeding convergence of these methods.

10.   Let (x_0, y_0) be a point in a region in which the head function satisfies the two-dimensional Laplace equation, and let C be a circle centered at (x_0, y_0). C can be of any radius r, large or small, as long as it and its interior are completely contained in the portion of the aquifer to which Laplace's equation applies. Show that the head value at the center of the circle, $h(x_0, y_0)$, is simply the average of the head values all around C, represented by the line integral

$$\frac{1}{2\pi r} \oint_C h(x, y).$$

This property is called the *mean value property of harmonic functions*. [Hint: show that this integral is actually independent of r, from which the result then follows easily.]

We write the given integral as a function of r and then show that its derivative with respect to r is 0, making it a constant. This involves parametrizing the specified circle to rewrite the integral as:

$$m(r) \equiv \frac{1}{2\pi r} \oint_C h(x, y)$$

$$= \frac{1}{2\pi r} \int_0^{2\pi} h(x_0 + r\cos\theta, y_0 + r\sin\theta) r\, d\theta$$

$$= \frac{1}{2\pi} \int_0^{2\pi} h(x_0 + r\cos\theta, y_0 + r\sin\theta) d\theta.$$

Note that in the third integral there is no division by r and, hence, this functional form applies to $r = 0$ as well and shows the differentiability of the function there based on the smoothness properties of h. Now we take the derivative of the function $m(r)$, and then apply Green's theorem (the divergence theorem in two dimensions) and the fact that h satisfies Laplace's equation:

$$m'(r) = \frac{1}{2\pi} \int_0^{2\pi} \left[h_x(x_0 + r\cos\theta, y_0 + r\sin\theta)\cos\theta + h_y(x_0 + r\cos\theta, y_0 + r\sin\theta)\sin\theta \right] d\theta$$

$$= \frac{1}{2\pi} \int_0^{2\pi} [\nabla h \cdot n] d\theta$$

$$= \frac{1}{2\pi} \int\int div(\nabla h)$$

$$= \frac{1}{2\pi} \int\int (h_{xx} + h_{yy})$$

$$= 0.$$

In the above, n is the unit outward normal to the circle, which we recognized as $(\cos\theta, \sin\theta)$, and the double integral is over the circle and its interior. Since the function $m(r)$ is constant, we need only find its value at some single point. At $r = 0$, its value is clearly $h(x_0, y_0)$, so that must be its value for all r, which completes the argument.

The result of this exercise can be seen to imply that a harmonic function defined on a closed disk must assume its maximum on the boundary, for if there were an interior maximum point, a circle and disk could be constructed around it in which this principle would be violated (unless the function were constant). This is the "maximum principle" for harmonic functions and is easily extendible to more general regions. From this it follows that if two harmonic functions have the same boundary values on such a region, they must be identical, for their difference is a harmonic function and its maximum would have to be 0. Thus one begins along these lines to pursue the uniqueness and existence of solutions under various types of boundary conditions.

5.9 Guide to Further Information

(No exercises.)

6

Additional Topics in Air Modeling and Diffusion Processes

6.1 Using Calculus to Obtain Further Information from the Diffusion Equation

1. **Without performing any numerical calculations but basing your answer solely on your physical intuition, what would be the general shape of the graph of the function just described, namely, C as a function of t for an arbitrary fixed value of x?**

    As long as $x \neq 0$, the point of initial mass injection, it should start at 0, rise to a peak, and then decay away to 0. (For example, see the second graph in Figure 3-13 in Chapter 3 of the text.) If on the other hand $x = 0$, then the concentration is not defined at 0, but it decays from $+\infty$ to 0 over the positive t axis.

2. **Does the general shape of the graph that you drew in connection with Exercise 1 hold for every single x value, or can you find one or more x values where the shape of the graph is qualitatively different? Once again, do this problem simply on the basis of your physical intuition.**

    As in Exercise 1, the behavior differs significantly between the cases $x \neq 0$ and $x = 0$.

3. **Now consider the one-dimensional diffusion situation for the following parameter set: $M = 200$ grams, and $D = 40$ cm^2/sec. For the point corresponding to $x = 100$ cm, use the standard max-min methods from calculus to find the value of time t at which the concentration at that point reaches its maximum. What is this maximum concentration? (Be sure to specify your units.)**

    In this case the concentration may be treated simply as a function of time. We will take the derivative and set it equal to 0 to find the single critical point. Then we will show that this gives a local maximum and, in fact, an absolute maximum.

    $$C = \frac{M}{\sqrt{4\pi Dt}} e^{-\frac{x^2}{4Dt}}, \qquad t > 0$$

    $$C_t = \frac{M}{\sqrt{4\pi Dt}} e^{-\frac{x^2}{4Dt}} \left(\frac{x^2}{4Dt^2} - \frac{1}{2t} \right) = \frac{M}{\sqrt{4\pi Dt}} e^{-\frac{x^2}{4Dt}} \cdot \frac{1}{2t} \cdot \left(\frac{x^2}{2Dt} - 1 \right) = 0$$

    $$x^2 = 2Dt$$

    $$t = \frac{x^2}{2D} = \frac{(100 \text{ cm})^2}{2 \cdot 40 \frac{\text{cm}^2}{\text{sec}}} = \frac{10000}{80} \text{ sec} = 125 \text{ sec.}$$

    By looking at the only potentially negative factor in the expression for the derivative, namely, $\frac{x^2}{2Dt} - 1$, it is clear that the derivative is positive for smaller t values than the critical value and negative for larger t values. Thus $t = 125$ must yield both a local and an absolute maximum concentration at this particular point $x = 100$ cm.

4.   Consider the general one-dimensional diffusion equation. For an arbitrary but fixed value of x, can you find a general expression for the time t at which the concentration at x will temporarily achieve its peak value? (The answer to this should obviously depend on x, as well as on one or more other parameters.)

This was actually worked out as part of the previous solution:

$$t = \frac{x^2}{2D}.$$

5.   Explain by physical reasoning why it was reasonable that the answer to the previous problem did not depend on the mass M.

The concentration is proportional to M at every location and time, since if you change the amount of original mass injected into the system, it still spreads out in the same proportions. (This fact can be seen in the diffusion equation itself.) Therefore, for a fixed location, if the concentration is greater at one time than at another, the same would be true for any other value of M. Therefore, the time giving the maximum value at this location would also not be affected by the particular value of M. Hence, its analytic determination, as in the previous exercise, should not depend on M.

6.   Show mathematically that the concentration value at the point corresponding to $x = 0$ is always decreasing with time, and then explain on physical grounds why this is precisely the result you would have expected.

For the point $x = 0$, the concentration function has meaning only for $t > 0$, and, in fact, it reduces to the expression

$$C = \frac{M}{\sqrt{4\pi Dt}}.$$

As t increases, so too does the denominator, making the fraction decrease towards 0. You would expect this on physical grounds because there is a net diffusion of material away from the injection point, $x = 0$, and so the concentration there should be decreasing.

7.   As described above, consider t to be a fixed point in time so that the concentration C may be regarded as a function of the single variable x. Demonstrate mathematically that this function has a relative (or local) maximum at the point $x = 0$.

Now we keep time fixed and look at C as a function of x, taking its derivative, setting it equal to 0, and solving for x:

$$C = \frac{M}{\sqrt{4\pi Dt}} e^{-\frac{x^2}{4Dt}}$$

$$C_x = \frac{M}{\sqrt{4\pi Dt}} e^{-\frac{x^2}{4Dt}} \cdot \left(\frac{-2x}{4Dt} \right) = 0$$

$$x = 0.$$

Thus there is a single critical point at $x = 0$, and consideration of the sign of the expression for C_x shows that this derivative is positive for negative x values and negative for positive x values. Therefore we have a local and absolute maximum at $x = 0$. This problem could be done without the use of derivatives by applying knowledge of the exponential function, which in this case always has a negative exponent for any $x \neq 0$, and hence is always less than one, the value when $x = 0$.

8.   To continue with the previous situation, demonstrate mathematically that the point $x = 0$ also corresponds to an absolute maximum.

This was included as part of the previous solution.

9. Suppose once again that you have a one-dimensional diffusion apparatus of the type discussed previously, and you are trying to plan an experiment to let you determine the value of the diffusion coefficient D. You plan to inject one gram of solute into the diffusion tube at the initial time. Assume that you have the ability to measure the concentration of diffusing material at any point and at any time, but you are permitted to carry out only one such measurement. Using such a measurement and assuming that all experimental errors are negligible, would you have enough information to determine D? Explain your answer. If your answer is "yes," illustrate with a numerical example. If your answer is "no," illustrate with a numerical counterexample.

No, you do not have quite enough information. In general, every location except $x = 0$ experiences most concentrations twice—once as the concentration is increasing at the point and once as it is decreasing. Only if you happened by accident to catch the concentration at the point precisely at the moment when it was peaking would there be only one possible value for D.

The development of a numerical example provides a good illustration of how solution techniques like Newton's method come in handy for nonlinear equations (although this could also be done by just repetitive numerical experiments). Our example will be based on the example that is sketched in Figure 3-13 in the text. Here $M = 5$, and suppose that we find by measurement that $C(2, 10) = .518884$, that value calculated in the table included in that figure. We know by reasoning backwards that one possible value of the diffusion coefficient is 0.1, the value that the figure is based on. This value gives us the D that would cause this concentration to appear on the uprising part of the concentration curve. However, there must be some larger D, meaning a faster diffusion process, that would move the peak through sooner and let us observe this concentration on the decreasing part of the concentration curve. To find this value and thereby to demonstrate this situation, note that for the given values of all the variables and parameters, we have:

$$C = \frac{M}{\sqrt{4\pi Dt}} e^{-\frac{x^2}{4Dt}}$$

$$.518884 = \frac{5}{\sqrt{4\pi D \cdot 10}} e^{-\frac{2^2}{4D \cdot 10}} = \frac{.446031}{\sqrt{D}} e^{\frac{-1}{10D}}$$

$$0 = 1.163336\sqrt{D} - e^{\frac{-1}{10D}} \equiv f(D).$$

This last equation defines a function $f(D)$ to which Newton's method can be applied. Most simple calculus computer programs implement this method, or it can be programmed into a spreadsheet. (Many spreadsheets also have equation solvers built in.) In any case, this function has the graph shown in Figure 6-A, which shows the known solution, $D = .1$, as well as the new larger one. The latter is found to have the value $D = .492$.

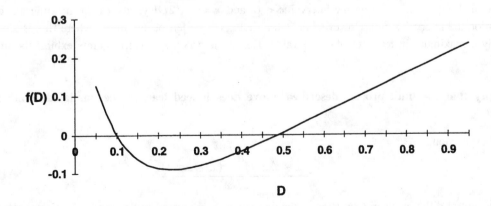

FIGURE 6-A
Graph of function $f(D)$ from solution to Exercise 9

10.   For the one-dimensional diffusion situation, is the flux the same at all locations, if you consider them all at the same fixed instant in time, or does it vary from location to location? Provide two distinct lines of reasoning to answer this question: one based on your physical intuition as applied to this situation, and the other based on mathematical analysis of the appropriate equation.

No, the flux is different, in general, at different points. For example, early in the process, the flux is very small far out along the diffusion tube, where there is practically no material and no gradient, although it is very large near the injection point, where there is rapid initial diffusion due to the steep concentration gradient. Mathematically, the flux is proportional to the concentration gradient (this is the diffusion principle), and the latter has been found earlier to be given by

$$C_x = \frac{M}{\sqrt{4\pi Dt}} e^{-\frac{x^2}{4Dt}} \cdot \left(\frac{-2x}{4Dt} \right),$$

and this clearly varies with x.

11.   For a fixed time t, find any and all points of inflection of the concentration considered as a function of x. Give a physical interpretation of these x locations in terms of a special property that the flux has at those points.

Points of inflection are points where the concavity changes. Therefore we look for points where the second derivative with respect to x is 0:

$$C_x = \frac{M}{\sqrt{4\pi Dt}} e^{-\frac{x^2}{4Dt}} \cdot \left(\frac{-2x}{4Dt} \right)$$

$$C_{xx} = \frac{M}{\sqrt{4\pi Dt}} e^{-\frac{x^2}{4Dt}} \cdot \left(\frac{-2}{4Dt} \right) - \frac{2x}{4Dt} \cdot \frac{M}{\sqrt{4\pi Dt}} e^{-\frac{x^2}{4Dt}} \cdot \left(\frac{-2x}{4Dt} \right)$$

$$= \frac{M}{\sqrt{4\pi Dt}} e^{-\frac{x^2}{4Dt}} \cdot \left(\frac{-2}{4Dt} \right) \cdot \left[1 - \frac{x^2}{2Dt} \right] = 0$$

$$x = \pm\sqrt{2Dt}.$$

The factor in square brackets in this sequence is the only one that can change sign. Thus we see that the second derivative is positive outside the interval between the candidate points and negative between them. Thus each of the points $x = \pm\sqrt{2Dt}$ is an inflection point. You can see the change in concavity on a graph of a typical concentration curve, as in Figure 3-13 of the text.

The important physical property that we have at these points is that the flux is a maximum there. To see this, use the calculations above to conclude that $x = -\sqrt{2Dt}$ yields a local minimum and an absolute minimum of the flux (which equals $-KC_x$) on the interval $x < 0$, and $x = +\sqrt{2Dt}$ yields a local maximum and an absolute maximum on the interval $x > 0$. However, since the first is on the left of the origin, where flux is always negative, it is really a maximum in terms of absolute value. Hence these two symmetric points exhibit the maximum flux magnitude.

12.   Verify that the limit process described above does indeed lead to the partial derivative $\partial q/\partial x$ as indicated.

The expression

$$\frac{q(x - \Delta x, t) - q(x + \Delta x, t)}{2\Delta x}$$

is just the central difference approximation to the derivative, as discussed in Section 5.8, which contains a diagram. This is actually even a better approximation than the usual forward difference formula that is used in the basic definition of the derivative.

13. **[This is a particularly challenging exercise.] As anticipated earlier in the text, show that the one-dimensional diffusion equation does indeed satisfy the fundamental diffusion property and that the corresponding constant of proportionality is actually the diffusion coefficient D. (Hint: it may be helpful to use the relationship just derived above between the x partial derivative of q and the t partial derivative of C.)**

We know from the text discussion that the flux satisfies the equation $q_x = -C_t$, and therefore, by integrating from 0 to x, we obtain:

$$q(x, t) = q(x, t) - 0 = q(x, t) - q(0, t) = \int_0^x q_x \, dx = \int_0^x -C_t \, dx.$$

Therefore, we would be done if we could show that this last integral were just $-DC_x$, or, equivalently, that DC_x is an antiderivative of C_t and that it is the right one (i.e., it corresponds to the right constant of integration). The hard part is to show that it is an antiderivative of C_t. First we have to compute all the relevant derivatives: To begin, we calculate C_t using the product rule for derivatives:

$$
\begin{aligned}
C_t &= \frac{M}{\sqrt{4\pi Dt}} e^{-\frac{x^2}{4Dt}} \left(-\frac{x^2}{4D}\right)\left(\frac{-1}{t^2}\right) + e^{-\frac{x^2}{4Dt}} \frac{M}{\sqrt{4\pi D}} \left(-\frac{1}{2} t^{-\frac{3}{2}}\right) \\
&= \frac{M}{\sqrt{4\pi Dt}} e^{-\frac{x^2}{4Dt}} \left[\frac{x^2}{4Dt^2} - \frac{1}{2t}\right] \\
&= \frac{M}{\sqrt{4\pi Dt}} e^{-\frac{x^2}{4Dt}} \left[\frac{x^2 - 2Dt}{4Dt^2}\right].
\end{aligned}
$$

Now proceeding to the x derivatives, we first have:

$$C_x = \frac{M}{\sqrt{4\pi Dt}} e^{-\frac{x^2}{4Dt}} \cdot \frac{-2x}{4Dt}$$

from which we get the second derivative by the product rule:

$$
\begin{aligned}
C_{xx} &= \frac{M}{\sqrt{4\pi Dt}} \left[e^{-\frac{x^2}{4Dt}} \cdot \frac{-2}{4Dt} + \left(\frac{-2x}{4Dt}\right) e^{-\frac{x^2}{4Dt}} \cdot \frac{-2x}{4Dt} \right] \\
&= \frac{M}{\sqrt{4\pi Dt}} e^{-\frac{x^2}{4Dt}} \left[\frac{-1}{2Dt} + \frac{4x^2}{(4Dt)^2} \right] \\
&= \frac{M}{\sqrt{4\pi Dt}} e^{-\frac{x^2}{4Dt}} \left[\frac{-2Dt + x^2}{4D^2 t^2} \right].
\end{aligned}
$$

Now we can see by inspection of terms that

$$\frac{\partial}{\partial x} DC_x = DC_{xx} = C_t.$$

Hence, using this antiderivative in the fundamental theorem of calculus:

$$q(x, t) = \int_0^x -C_t \, dx = -DC_x \Big|_0^x = -DC_x(x, t) + DC_x(0, t) = -DC_x(x, t)$$

which completes the solution. You can see that the term $DC_x(0, t)$ is 0 either by plugging $x = 0$ into the expression for this derivative or noting that the bell-shaped curve giving the graph of C has a horizontal tangent at this point.

14. **It is not difficult to attach intuitive meaning to the statement "diffusion occurs faster in experiment A than in experiment B." We would think of two one-dimensional diffusion tubes A and B where the combination of solute and substrate in tube A is such that the solute tends to spread out through the tube faster than in the corresponding tube B. We also assume that we begin with the same mass M in each case. Which of the following are logically equivalent to the above statement?**

a) For every pair of values of x and t, the magnitude of the flux value in A is larger than the corresponding value in B.

Not equivalent. For example, consider an experiment where the diffusion in A is extremely rapid, almost instantaneous, and the diffusion in B is moderately slow. A may "spend itself" so very fast that it is reduced to a very flat curve with negligible values. This would also then imply very tiny fluxes, even back near the origin where the flux in B may still be considerable.

b)   The diffusion coefficient for A is smaller than the diffusion coefficient for B.

Not equivalent. In fact, just the opposite is true. A higher D value means a higher flux for the same concentration gradient, since D is the actual constant of proportionality in the diffusion principle. In other words, faster diffusion is equivalent to the condition of a higher D, a fact that we will use in the next two parts.

c)   For every time t greater than 0, the magnitude of the drop rate in the concentration at the origin, given by the time derivative $C_t(0, t)$, is greater for A than for B.

Not equivalent. Since this derivative at the point $x = 0$ is given by

$$C_t = \frac{d}{dt}\left(\frac{M}{\sqrt{4\pi Dt}}\right) = \frac{M}{\sqrt{4\pi D}} \cdot \frac{-1}{2t^{3/2}},$$

if we fix the value of time, a larger D value corresponds to a *smaller* magnitude of C_t, rather than a larger one. (This may be surprising, but see the next part.)

d)   For every time t greater than 0, the magnitude of the concentration at the origin, given by $C(0,t)$, is less for A than for B.

Equivalent. Since the concentration at the origin is given by $C = \dfrac{M}{\sqrt{4\pi Dt}}$, for any fixed t, a larger D value will imply a lower concentration. This creates a paradoxical situation in combination with the previous part, for we have in the A situation a lower fall rate in C and yet always a lower value too. This is possible because C has a singularity at $t = 0$.

e)   For every time t greater than 0, the positive point of maximum flux in A is farther to the right than the corresponding point of maximum flux in B.

Equivalent. This point was seen earlier to be given by $x = +\sqrt{2Dt}$, so a larger D value will cause it to be farther to the right.

15.   We have been talking loosely and intuitively about the concept of a quantity of mass M injected at the point $x = 0$ at the initial time $t = 0$. There is a subtle logical problem associated with this concept. Can you identify this problem, and can you suggest a mathematical approach to this concept that shows promise in getting around the problem?

The idea of a mass M concentrated at one point does not, in itself, pose a major logical problem, although it would be impossible to implement in a real experiment. Nevertheless, M could consist of a single molecule, and we are not unaccustomed to thinking of such a molecule as being located at a point, even if it is still an idealization. The real problem is to describe how the system starts to evolve at $t = 0$. Since the concentration is not well defined at this time (it would be infinite at $x = 0$ and 0 elsewhere), we can't talk about the concentration gradient or the diffusion principle, so we can't seem to get "off the mark" into the regime where these concepts make sense.

A potentially reasonable definition could go as follows, although it too has limitations. For any number δ, which we will we shortly let approach 0, consider a diffusion tube experiment in which we begin at time $t = 0$ with the mass M uniformly distributed in the tube between $-\delta$ and $+\delta$. This is more physically meaningful than the idealized concept of injecting mass instantaneously at a single point. The resulting concentration function, denoted $C_\delta(x,t)$, makes sense for all x and for all $t \geq 0$. Of course it may not have the special form of our one-dimensional diffusion equation because the starting condition is quite different. If we now take the limit as $\delta \to 0$ of the concentration $C_\delta(x,t)$ at each (x,t) point, the limiting value, which we could call $C(x,t)$, may be regarded as the concentration value for the idealized problem. Note that the limit does not exist at $x = 0$ because the values increase without bound.

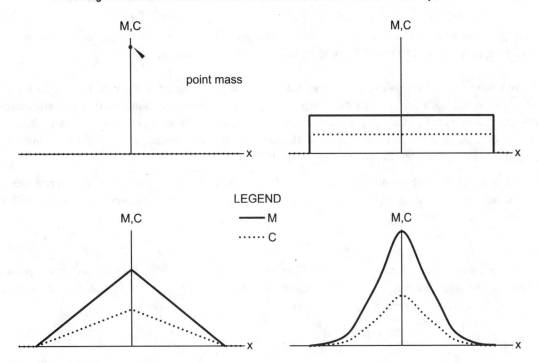

FIGURE 6-B

Possible corresponding mass and concentration functions for Exercise 15 (not to scale)

This represents a reasonable approach but, unfortunately, it has its own problems. For example, at time $t = 0$, what would be the flux values at the points $x = \pm\delta$? Since $C_\delta(x, 0)$ is easily seen not to be continuous there, it is certainly not differentiable, and so we could hardly apply the diffusion principle to get the system "off the mark" again. This approach is shown in the second graph of Figure 6-B.

This shortcoming is easier to deal with. As a next step we could start with a triangular mass distribution, as shown in the third section of Figure 6-B. That is, the total mass within the triangle would be M, but it would be most concentrated at the center and then taper out linearly to 0 at the points $x = \pm\delta$. This gives a continuous concentration function $C_\delta(x, 0)$, but still not a differentiable one, so the initial flux at $x = \pm\delta$ would still be undetermined. But now the final strategy is emerging. We want a very smooth initial mass distribution, like the one shown in the fourth part of the figure. The total mass would still be M, but the concentration function would be such that the gradient $\dfrac{\partial C_\delta(x, 0)}{\partial x}$ would exist at every point, and thus the diffusion principle would be enough to define how the system would get going. In fact, we could easily find such a function so that the second partial derivative $\dfrac{\partial^2 C_\delta(x, 0)}{\partial x^2}$ would also exist. An example of such a function would be $\dfrac{15M}{16\delta^5}(x - \delta)^2(x + \delta)^2$.

With this family of functions, we could now define the solution to the original idealized problem as the limit given earlier, but we would still not be done. For example, if someone else came along and defined the idealized solution with the same line of logic, but that person used a similar although not quite the same family of functions for the initial mass distribution, would the same final answer result? This would need to be determined in the affirmative in order for the present line of reasoning to be truly appropriate. In fact, to anticipate the representation of the total mass in any interval as an integral (see the next section of the text), we might also consider mass distributions that may even always be actually greater than 0 for every x and for every $t > 0$, as long as they satisfy the limit condition that for any fixed δ,

$$M = \lim_{t \to 0} \int_{-\delta}^{\delta} C(x, t)\, dx,$$

which says that an arbitrarily high fraction of the mass is concentrated in any fixed interval, however small, as long as t is sufficiently small. (Our one-dimensional diffusion equation would satisfy this condition.) Although we shall not pursue this issue here, it does lead to some very useful generalizations of differential equations and boundary conditions, an appropriate mathematical framework being the theory of distributions.

16.   In the context of the preceding discussion, imagine that you divide the interval from a to b up into n subintervals each of length Δx, so that $\Delta x = (b-a)/n$. Use the concentration function to write expressions for the approximate value of the total mass contained with the first subinterval, the second subinterval, etc. Then use the definition of the definite integral as the limit of sums to demonstrate that the total mass between a and b can be represented by the integral given in the text.

Our definition of the concentration at a point as a limit would imply that for any such subinterval, the concentration at the midpoint should be approximately equal to the total mass in the subinterval divided by the length of the subinterval. Equivalently,

$$\text{mass in subinterval} \approx C(c_i, t)\Delta x$$

where c_i is the midpoint of the ith subinterval. (The statement would also be true for any other point within the subinterval; the midpoint was taken only to correspond to the exact way it was stated in the definition of concentration.) Now we add the equations for all i and take the limit as $n \to \infty$:

$$\text{total mass} \approx \sum_i C(c_i, t)\Delta x$$

$$\text{total mass} = \lim_{n\to\infty} \sum_i C(c_i)\Delta x = \int_a^b C(x,t)\, dx$$

where the last step comes from the basic definition of the definite integral as a limit of Riemann sums.

17.   Consider a specific one-dimensional diffusion problem in which three ounces of material are injected at the center of a long thin tube for diffusion within the substrate, and assume that the diffusion constant D has the value 5 in^2/sec. At the time given by $t = 80$ sec, find the total amount of mass in that portion of the tube between the points corresponding to $x = 3$ and $x = 25$. If you encounter some portion of your calculation where you cannot find an exact value, use tools at your disposal to approximate the numerical value.

The concentration function for this case has the form

$$C = \frac{M}{\sqrt{4\pi D t}}e^{-\frac{x^2}{4Dt}} = \frac{3}{\sqrt{4\pi \cdot 5 \cdot 80}}e^{-\frac{x^2}{4\cdot 5\cdot 80}} = \frac{3}{\sqrt{1600\pi}}e^{-\frac{x^2}{1600}},$$

and while it cannot be integrated in closed form, we can use a standard numerical method, available in most calculus packages, to determine that the answer is .808 ounces.

18.   Find the value of the following improper integral:

$$\int_{-\infty}^{\infty} \frac{M}{\sqrt{4\pi D t}}e^{-\frac{x^2}{4Dt}}\, dx$$

(Hint: there is both an easy way and a very hard way to answer this question.)

The answer must be M because this is the integral of the concentration over the entire interval, so you must get the total mass. (The integral is actually an improper integral, not a Riemann integral as treated in Exercise 16, so technically we should be looking at it as a limit. But that limit is clearly going to be the total mass in the system.)

19.   Consider the two-dimensional diffusion equation in the situation where the diffusion constants D_1 and D_2 are equal to a single value D. Show that the value of the concentration at any point can be represented as a function of a single new spatial variable. (Hint: you should be able to anticipate in advance what this variable is even before you apply some simple algebraic manipulations to obtain the required form.)

Since the diffusion constant is the same in all directions, you should expect this single variable to be the distance r from the point of mass injection. This is exactly what results from the algebraic manipulation:

$$C = \frac{M}{4\pi t \sqrt{D^2}} e^{\left(-\frac{x^2}{4Dt} - \frac{y^2}{4Dt}\right)} = \frac{M}{4\pi t D} e^{-\frac{x^2+y^2}{4Dt}} = \frac{M}{4\pi t D} e^{-\frac{r^2}{4Dt}}.$$

20. **For this same two-dimensional diffusion problem with equal diffusion coefficients, describe the family of geometric curves in the x, y-plane along each of which the resulting values of concentration will be constant.**

These are lines of constant distance r from a fixed point, so they are circles.

21. **For the two-dimensional diffusion situation in which the diffusion constants are different in the x- and y-directions, describe the precise class of curves in the x, y-plane that represent points of constant concentration. These are sometimes called isopleths, contour lines (as in the case of ground water), or level lines.**

These are ellipses, since they are described by equations of the form:

$$\frac{x^2}{4D_1 t} + \frac{y^2}{4D_2 t} = \text{constant.}$$

22. **[Involves multiple integration.] For the two-dimensional diffusion situation in which the diffusion constants are different in the x- and y-directions, what should the value be of the double integral of the two-dimensional diffusion equation over the entire plane? Verify, using the techniques of integration for such integrals, that you do get the value that you expect on physical grounds. (Hint: this problem is not so hard as it may sound.)**

The value should be M since the integral represents the total mass in the system. It turns out to be easy because it can be done as an iterated integral:

$$\iint_{R^2} \frac{M}{4\pi t \sqrt{D_1 D_2}} e^{\left(-\frac{x^2}{4D_1 t} - \frac{y^2}{4D_2 t}\right)} dA = \int_{-\infty}^{\infty} \int_{-\infty}^{\infty} \frac{M}{4\pi t \sqrt{D_1 D_2}} e^{\left(-\frac{x^2}{4D_1 t} - \frac{y^2}{4D_2 t}\right)} dx \, dy$$

$$= \int_{-\infty}^{\infty} \frac{1}{\sqrt{4\pi D_2 t}} e^{-\frac{y^2}{4D_2 t}} \left[\int_{-\infty}^{\infty} \frac{M}{\sqrt{4\pi D_1 t}} e^{-\frac{x^2}{4D_1 t}} dx\right] dy$$

$$= \int_{-\infty}^{\infty} \frac{M}{\sqrt{4\pi D_2 t}} e^{-\frac{y^2}{4D_2 t}} dy$$

$$= M.$$

The successive one-dimensional integrals each have the value M because they are essentially the same integral as in Exercise 18.

23. **[Partly involves multiple integration, but the equation should be able to be guessed at without one's being experienced with such integrals.] Write down what you would expect to be the obvious form for the *three-dimensional diffusion equation* based on analogy with the equations for the one- and two-dimensional cases. Assume the possibility of different diffusion constants in each of the three principal directions. Use the integration constraint, of the type that you investigated for the two-dimensional case in the previous exercise, to determine the correct constant(s) for the three-dimensional diffusion equation.**

The form would be expected to be

$$C = \frac{M}{(4\pi t)^{3/2} \sqrt{D_1 D_2 D_3}} e^{-\frac{x^2}{4D_1 t} - \frac{y^2}{4D_2 t} - \frac{z^2}{4D_3 t}}.$$

The only real area where there might be a question has to do with the factor $(4\pi t)^{3/2}$, which follows the apparent pattern of multiplying by $\dfrac{1}{\sqrt{4\pi t}}$ each time another dimension is added. We verify the correctness of this factor by using three iterated integrals to integrate the concentration over 3-space, analogous to what was done for 2-space in the previous solution:

$$\iiint_{R^3} \frac{M}{(4\pi t)^{3/2}\sqrt{D_1 D_2 D_3}} e^{\left(-\frac{x^2}{4D_1 t}-\frac{y^2}{4D_2 t}-\frac{z^2}{4D_3 t}\right)}\, dV$$

$$= \int_{-\infty}^{\infty}\int_{-\infty}^{\infty}\int_{-\infty}^{\infty} \frac{M}{(4\pi t)^{3/2}\sqrt{D_1 D_2 D_3}} e^{\left(-\frac{x^2}{4D_1 t}-\frac{y^2}{4D_2 t}-\frac{z^2}{4D_3 t}\right)}\, dx\, dy\, dz$$

$$= \int_{-\infty}^{\infty} \frac{1}{\sqrt{4\pi D_3 t}} e^{-\frac{z^2}{4D_3 t}} \left\{ \int_{-\infty}^{\infty} \frac{1}{\sqrt{4\pi D_2 t}} e^{-\frac{y^2}{4D_2 t}} \left[\int_{-\infty}^{\infty} \frac{M}{\sqrt{4\pi D_1 t}} e^{-\frac{x^2}{4D_1 t}}\, dx \right] dy \right\} dz$$

$$= M$$

where the individual integrals turn out successively to have the value M just as in the previous solution.

6.2 Relation Between the Diffusion Equation and the Gaussian or Normal Distribution from Statistics

1.   Compare the corresponding terms from the normal distribution and the one-dimensional diffusion equation to identify the correspondence between the parameters and variables in one equation and those in the other.

By observation, the correspondence is as follows: $\mu = 0$ and $\sigma = \sqrt{2Dt}$. Note that with the second, one has to verify that it does indeed agree with the two locations where these terms show up.

2.   Based on probability theory, find the value of the integral

$$\int_{-\infty}^{\infty} \frac{1}{\sqrt{2\pi}\sigma} e^{-\frac{1}{2}\left(\frac{x-\mu}{\sigma}\right)^2}\, dx.$$

Describe the relationship between this result and the answer to a very analogous exercise in the previous section.

The value of the integral is 1, since it is a probability function, and hence the integral over the entire space must be 1. This is analogous to the integration of the one-dimensional diffusion equation in Exercise 18 in the previous section. In fact, you could use either one to derive the other. For example, if you convert the integral in this problem to the diffusion notation, using the correspondence in the previous exercise, you can obtain the result of Exercise 18 as follows:

$$M = M\int_{-\infty}^{\infty} \frac{1}{\sqrt{2\pi}\sigma} e^{-\frac{1}{2}\left(\frac{x-\mu}{\sigma}\right)^2}\, dx = M\int_{-\infty}^{\infty} \frac{1}{\sqrt{4\pi Dt}} e^{-\frac{x^2}{4Dt}}\, dx = \int_{-\infty}^{\infty} \frac{M}{\sqrt{4\pi Dt}} e^{-\frac{x^2}{4Dt}}\, dx.$$

3.   Give a heuristic (i.e., intuitive) explanation, referring to the comparison developed within Exercise 1, above, of why the counterpart for the standard deviation σ from the normal distribution does indeed appear reasonable. Be sure to consider both components, D and t, of this counterpart in your explanation.

The correspondence was $\sigma = \sqrt{2Dt}$. Since the standard deviation σ is a measure of the amount of spread of the normal distribution, it is indeed reasonable that this should increase with both the diffusion coefficient D (faster diffusion) and time t (more time for spreading).

4. [This problem presumes that the reader has studied basic multivariable calculus.] Prove from the basic principles of calculus that the integral of the normal distribution from $-\infty$ to $+\infty$ is equal to 1, as asserted in terms of its probability interpretation above. (Hint: this is a standard problem from multivariable calculus wherein it is usually suggested that one square the given integral, using two different letters for the dummy variables in the two integrals, and then convert the result to a double integral written in polar coordinates.)

First we convert the integral of the normal distribution to the "standard normal distribution" which has $\mu = 0$ and $\sigma = 1$. This is done by creating a new variable of integration $z = \dfrac{x - \mu}{\sigma}$, in terms of which the integral becomes $\displaystyle\int_{-\infty}^{\infty} \dfrac{1}{\sqrt{2\pi}} e^{-\frac{z^2}{2}} \, dz$. Just to keep with the familiar spatial variables x and y, we will actually rewrite this with x as the dummy variable of integration. So now we have:

$$\left(\int_{-\infty}^{\infty} \frac{1}{\sqrt{2\pi}} e^{-\frac{x^2}{2}} \, dx\right)^2 = \left(\int_{-\infty}^{\infty} \frac{1}{\sqrt{2\pi}} e^{-\frac{x^2}{2}} \, dx\right)\left(\int_{-\infty}^{\infty} \frac{1}{\sqrt{2\pi}} e^{-\frac{y^2}{2}} \, dy\right)$$

$$= \int_{-\infty}^{\infty}\int_{-\infty}^{\infty} \frac{1}{2\pi} e^{-\frac{x^2+y^2}{2}} \, dx \, dy$$

$$= \iint_{R^2} \frac{1}{2\pi} e^{-\frac{x^2+y^2}{2}} \, dA$$

$$= \int_{0}^{\infty}\int_{0}^{2\pi} \frac{1}{2\pi} e^{-\frac{r^2}{2}} \, r \, d\theta \, dr$$

$$= \int_{0}^{\infty} r e^{-\frac{r^2}{2}} \, dr = -e^{-\frac{r^2}{2}}\Big|_{0}^{\infty} = 1.$$

The use of the infinity symbol in evaluating the integral is of course technically an abbreviation for taking the limit as the upper bound approaches infinity. Thus the original integral, which must be positive since the integrand is positive, must also have the value 1.

5. When the binomial distribution for large values of n is approximated by the corresponding normal distribution as described above, what are the values of the parameters σ and μ needed to specify the normal distribution? (You do not have to derive the answer. The value for μ should be easy to see, but the value for σ may be a bit more difficult if you have not come across it before. You may wish to consult a reference to find it.)

The corresponding values are $\mu = n/2$ and $\sigma = \sqrt{n/4}$. The latter one is not obvious but is simply the standard deviation of the binomial distribution; it can be determined by applying the definition of standard deviation or variance to this distribution.

6. The probability function for the binomial distribution with equally likely outcomes has been given in the text above. Write the corresponding probability function for a binomial distribution where the outcomes are not necessarily equally likely, so that they must be characterized by probabilities p and $1 - p$.

For k successes in n trials where the probability of a success is p,

$$\text{Prob}(k \text{ successes}) = \binom{n}{k}(p)^k(1-p)^{n-k} = \frac{n!}{k!(n-k)!}(p)^k(1-p)^{n-k}.$$

7. When the value of p is not equal to 1/2, the histogram that you obtain when graphing the probabilities of various possible outcomes is no longer symmetric, unlike the histograms shown in the figures in this section. Nevertheless, the normal distribution is always symmetric. Does the binomial distribution converge to the normal distribution only for the case of equally likely outcomes, or does it converge in the more general

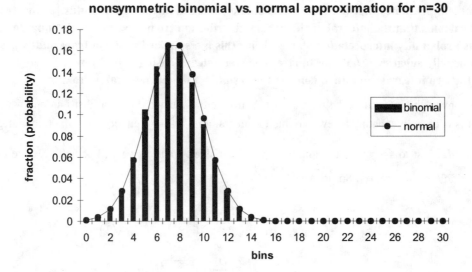

FIGURE 6-C

Comparison of binomial distribution and normal approximation in a nonsymmetric case (Exercise 7)

case? Explain your answer, perhaps illustrating with a numerical experiment. (Hint: you may wish to consult some statistics reference material to answer this question.)

Yes, the approximation is still valid because the normal curve is still an asymptotic approximation to the binomial distribution. An example is shown in Figure 6-C where the binomial distribution for $n = 30$ and $p = .25$ is shown together with its normal approximation. The approximating normal distribution is centered on the mean of the binomial distribution. Although you can see the asymmetry in the binomial even in the vicinity of its peak, the approximation seems quite good in this fairly representative example. (A rough "rule of thumb" often applied in statistics is that the normal approximation to the binomial distribution can be used when both $np \geq 5$ and $n(1 - p) \geq 5$.)

8. [Difficult problem.] As has been asserted in the text, for large values of n, the binomial distribution follows very closely the shape of the bell-shaped curve given by the normal distribution. Prove this fact mathematically for the case of equally likely outcomes, as applied in the text. (Hint: the key fact you will need is Stirling's formula, which gives an approximation to factorials in the form: $n! \approx \dfrac{n^n}{e^n}\sqrt{2\pi n}$. The approximation is an asymptotic approximation, meaning that as $n \to \infty$, the ratio between the exact and approximate values approaches 1. You may also want to keep in mind the following limit for e: $e = \lim_{n\to\infty}(1 + \frac{1}{n})^n$, which can also be written: $e = \lim_{\varepsilon \to 0}(1 + \varepsilon)^{1/\varepsilon}$.)

Even though this problem is labeled as "difficult," it is one that a good student should be able to work out without consulting a reference, and it should remove the general mystery that often surrounds this approximation in students' eyes. The limits are slightly more complicated in the general nonsymmetric case, but the ideas are basically the same. The derivation given here appears to the author to be more straightforward and motivated than is often encountered in statistics references.

First we have to find the specific comparison that should be applicable and then we have to verify it. For the former, it will be useful to follow the logical sequence illustrated in Figure 6-D at the same time that the details are discussed here. We begin with the probability function for the binomial distribution:

$$p(x) = \binom{n}{x}\left(\frac{1}{2}\right)^n, \quad \text{for } x = 0, 1, 2, \ldots, n.$$

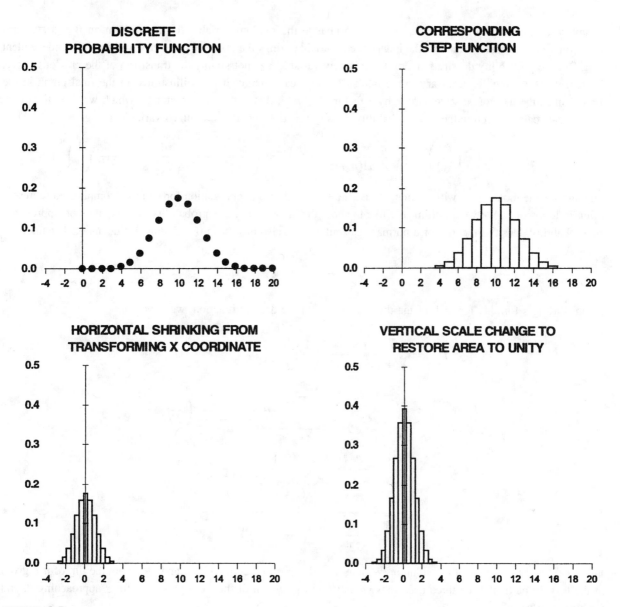

FIGURE 6-D

Transformation from a discrete to a corresponding continuous distribution (Exercise 8)

Now we want to transform this into a corresponding density function. We can do this simply by considering it as a step function with each step centered on one of the n discrete x values from 0 to n, and the resulting curve would indeed have the area under it equal to 1. However, the center value, around which the curve is symmetric, is located at $x = n/2$, and this will be migrating to infinity when we look at the limit as $n \to \infty$. Unfortunately, this would be quite inconvenient. Therefore, we will "normalize" our random variable by changing its x values to a new variable z given by

$$z = \frac{x - \mu}{\sigma} = \frac{x - \frac{n}{2}}{\sqrt{\frac{n}{4}}}.$$

If you are accustomed to such transformations, you may be even more accustomed to the general form

$$z = \frac{x - \mu}{\sigma} = \frac{x - np}{\sqrt{npq}}.$$

In any case, this transformation is easily seen to change the random variable to one with mean 0 and standard deviation 1. This transformation of the independent variable brings the point of symmetry to the origin, independent of n. But the probability density function for this new variable z is not simply the transform of the previous curve. A vertical "stretching" is necessary to reestablish a unit area underneath it, as illustrated in the final graph in the figure. Since the horizontal scale shrunk by a factor of $\sqrt{n/4}$, this is the same factor by which we need to stretch the vertical scale. Therefore the new probability density function for the normalized variable z is given by

$$f(z) = \sqrt{\frac{n}{4}} \binom{n}{x} \left(\frac{1}{2}\right)^n = \sqrt{\frac{n}{4}} \frac{n!}{x!(n-x)!} \left(\frac{1}{2}\right)^n = \sqrt{\frac{n}{4}} \frac{n!}{(\frac{n}{2}+z\sqrt{\frac{n}{4}})!(\frac{n}{2}-z\sqrt{\frac{n}{4}})!} \left(\frac{1}{2}\right)^n$$

although for the moment we will continue to use the symbol x as an abbreviation for its more complicated representation in terms of z. We would ultimately like to show that as $n \to \infty$, this probability density function approaches the probability density function for a normal distribution that also has mean 0 and standard deviation 1, namely:

$$\frac{1}{\sqrt{2\pi}} e^{-\frac{z^2}{2}}.$$

Using Stirling's formula for each of the three factorials in the density function, we have

$$f(z) = \sqrt{\frac{n}{4}} \frac{n!}{x!(n-x)!} \left(\frac{1}{2}\right)^n$$

$$\approx \sqrt{\frac{n}{4}} \frac{\dfrac{n^n}{e^n}\sqrt{2\pi n}}{\dfrac{x^x}{e^x}\sqrt{2\pi x}\dfrac{(n-x)^{n-x}}{e^{n-x}}\sqrt{2\pi(n-x)}} \left(\frac{1}{2}\right)^n$$

$$= \frac{1}{\sqrt{2\pi}} \cdot \frac{n^{n+1}}{2^{n+1}} \cdot \frac{1}{x^{x+\frac{1}{2}}(n-x)^{n-x+\frac{1}{2}}}$$

$$= \frac{1}{\sqrt{2\pi}} \cdot \frac{1}{\left(\dfrac{2x}{n}\right)^{x+\frac{1}{2}}\left(\dfrac{2(n-x)}{n}\right)^{n-x+\frac{1}{2}}}.$$

In the next and final stage of the solution, we convert all terms to the variable z, and we apply the limit for e given in the text. In particular, on three occasions we have an expression of the form $(1+\varepsilon)^\beta$, for ε approaching 0, and we replace it as follows:

$$(1+\varepsilon)^\beta = (1+\varepsilon)^{\frac{1}{\varepsilon}\cdot\varepsilon\beta} \to e^{\varepsilon\beta},$$

then proceeding to evaluate the limit of the numerator. This sequential limit-taking is valid as long as the limit of the exponent is finite.

So now we are ready for the final computation:

$$\lim_{n\to\infty} \frac{1}{\sqrt{2\pi}} \cdot \frac{n^{n+1}}{2^{n+1}} \cdot \frac{1}{x^{x+\frac{1}{2}}(n-x)^{n-x+\frac{1}{2}}}$$

$$= \frac{1}{\sqrt{2\pi}} \cdot \lim_{n\to\infty} \frac{1}{\left(1+\dfrac{z}{\sqrt{n}}\right)^{\frac{n}{2}+z\sqrt{\frac{n}{4}}+\frac{1}{2}}\left(1-\dfrac{z}{\sqrt{n}}\right)^{\frac{n}{2}-z\sqrt{\frac{n}{4}}+\frac{1}{2}}}$$

$$= \frac{1}{\sqrt{2\pi}} \cdot \lim_{n\to\infty} \frac{1}{\left(1-\dfrac{z^2}{n}\right)^{\frac{n}{2}}\left(1+\dfrac{z}{\sqrt{n}}\right)^{z\sqrt{\frac{n}{4}}}\left(1-\dfrac{z}{\sqrt{n}}\right)^{-z\sqrt{\frac{n}{4}}}\left(1-\dfrac{z^2}{n}\right)^{\frac{1}{2}}}$$

$$= \frac{1}{\sqrt{2\pi}} \cdot \lim_{n \to \infty} \frac{1}{\left(1 - \frac{z^2}{n}\right)^{-\frac{n}{z^2}\left(\frac{-z^2}{2}\right)} \left(1 + \frac{z}{\sqrt{n}}\right)^{\frac{\sqrt{n}}{z}\left(\frac{z^2}{2}\right)} \left(1 - \frac{z}{\sqrt{n}}\right)^{-\frac{\sqrt{n}}{z}\left(\frac{z^2}{2}\right)} \left(1 - \frac{z^2}{n}\right)^{\frac{1}{2}}}$$

$$= \frac{1}{\sqrt{2\pi}} \cdot \frac{1}{e^{-\frac{z^2}{2}} \cdot e^{+\frac{z^2}{2}} \cdot e^{+\frac{z^2}{2}} \cdot 1}$$

$$= \frac{1}{\sqrt{2\pi}} e^{-\frac{z^2}{2}}.$$

Therefore, by virtue of the fact that Stirling's formula is an asymptotic approximation, this final expression represents an asymptotic approximation to the binomial distribution.

Exercise 9.   Show that the product of a normal random variable and a constant is itself a normal random variable. (Hint: show that there are appropriate mean and standard deviation values so that the probability function for the new random variable fits the general form of a normal distribution.)

Starting with a random variable X with mean μ and standard deviation σ, we form a new random variable $Y = aX$ for some constant a. The basic question is: what probability density function can be used to determine Prob $\{c \le y \le d\}$ for any constants c and d? (We expect it to be the appropriate normal distribution.)

We therefore simply carry out the computation, changing the variable of integration:

$$\text{Prob}\,\{c \le y \le d\} = \text{Prob}\left\{\frac{c}{a} \le \frac{y}{a} \le \frac{d}{a}\right\} = \text{Prob}\left\{\frac{c}{a} \le x \le \frac{d}{a}\right\}$$

$$= \int_{c/a}^{d/a} \frac{1}{\sqrt{2\pi}\sigma} e^{-\frac{1}{2}\left(\frac{x-\mu}{\sigma}\right)^2} dx$$

$$= \int_{c/a}^{d/a} \frac{1}{\sqrt{2\pi}\sigma} e^{-\frac{1}{2}\left(\frac{ax-a\mu}{a\sigma}\right)^2} \frac{1}{a} d(ax)$$

$$= \int_c^d \frac{1}{\sqrt{2\pi}a\sigma} e^{-\frac{1}{2}\left(\frac{y-a\mu}{a\sigma}\right)^2} dy,$$

and this has precisely the form of a normal distribution with mean $a\mu$ and standard deviation $a\sigma$.

10.   Show that if X is a normal random variable, then so too is $aX + b$, where a and b are constants. (Exercise 9 was a special case of this.)

This is very similar to the previous problem:

$$\text{Prob}\{c \le y \le d\} = \text{Prob}\left\{\frac{c-b}{a} \le \frac{y-b}{a} \le \frac{d-b}{a}\right\} = \text{Prob}\left\{\frac{c-b}{a} \le x \le \frac{d-b}{a}\right\}$$

$$= \int_{(c-b)/a}^{(d-b)/a} \frac{1}{\sqrt{2\pi}\sigma} e^{-\frac{1}{2}\left(\frac{x-\mu}{\sigma}\right)^2} dx$$

$$= \int_{(c-b)/a}^{(d-b)/a} \frac{1}{\sqrt{2\pi}\sigma} e^{-\frac{1}{2}\left(\frac{ax+b-a\mu-b}{a\sigma}\right)^2} \frac{1}{a} d(ax + b)$$

$$= \int_c^d \frac{1}{\sqrt{2\pi}a\sigma} e^{-\frac{1}{2}\left(\frac{y-a\mu-b}{a\sigma}\right)^2} dy,$$

and this has precisely the form of a normal distribution with mean $a\mu + b$ and standard deviation $a\sigma$.

11.   Relate the conclusion of the previous exercise to the relationship between the following two random variables for the pinball device: X, **the original bin-number (e.g., number of steps to the right) described in Figure 6-5, considered as a normal random variable; and** Y_{net}, **the bin number based on the alternative bin numbering scheme in the same figure.**

$Y_{net} = 2(X - 2) = 2X - 4$, or for the case of n levels of pegs,

$$Y_{net} = 2\left(X - \frac{n}{2}\right) = 2X - n,$$

so in the general case the mean shifts by $\frac{n}{2}$ to a fixed value of 0 and the standard deviation is doubled.

12.   In the four-step sequence of equations above, leading to the expression for the concentration function $C(x)$ **as a limit, provide a more rigorous justification than in the text for going from the second to the fourth and final step. (Hint: you may wish to apply the average value concept for a function, or you may find the mean value theorem for integrals helpful.)**

By the mean value theorem on integrals (since the integrand is continuous),

$$\int_{x-\Delta x}^{x+\Delta x} \frac{1}{\sqrt{2\pi}\sigma} e^{\frac{1}{2}\left(\frac{x}{\sigma}\right)^2} dx = \frac{1}{\sqrt{2\pi}\sigma} e^{\frac{1}{2}\left(\frac{c}{\sigma}\right)^2} \cdot 2\Delta x$$

for some c within the small interval. Thus as $\Delta x \to 0$, it follows that $c \to x$, hence leading to the final expression in the sequence.

13.   In the expression for the concentration $C(x)$ **derived in this section, it was asserted in the text that multiplication through by a constant would convert everything from molecule units to mass units, as we are accustomed to in the diffusion equation. Demonstrate this process.**

$$m \text{ molecules} \times \frac{1 \text{ mass unit}}{n \text{ molecules}} = \frac{m}{n} \text{ mass units}$$

so the constant would be $1/n$, which of course varies with the individual material.

14.   [For readers with a strong statistics background.] Find a precise statement of the central limit theorem in a statistic reference, and use it to clarify the nature of the normal approximation that has been applied in this section. Explain also why some people say that "the central limit theorem rescues many people who apply statistical tests without checking that the distributions of their variables meet the assumptions of those tests."

If S_n is the sum of n independent, identically distributed, random variables, $Y_1, Y_2, \ldots, Y_n$, with common mean μ and standard deviation σ, then the normalized sums

$$Z_n = \frac{S_n - n\mu}{\sigma\sqrt{n}},$$

which have mean 0 and standard deviation 1, are "asymptotically normal" in the sense that their distribution functions $F_n(x)$, which represent the cumulative probability of an outcome less than or equal to x, satisfy the limiting condition

$$\lim_{n\to\infty} F_n(x) = \frac{1}{\sqrt{2\pi}} \int_{-\infty}^{x} e^{-\frac{s^2}{2}} ds,$$

the right-hand side being the distribution function for the standard normal distribution. (The distribution function is used in this typical formulation because the theorem covers more than just probability distributions with continuous probability density functions or continuous pointwise limits.)

With respect to the text's application, while the theorem is stated in terms of the normalized sums, once we know that they can be approximated by standard normal variables, it follows that the unnormalized versions, S_n, can be approximated by normal distributions with mean $n\mu$ and standard deviation $\sigma\sqrt{n}$, which were used in the text.

Many statistical tests involve taking multiple samples from a distribution and averaging the results. This is a two-step process: summing (the complex portion) and then dividing by the number of samples (the simple portion, akin to multiplying by a constant, which preserves normality). The central limit theorem tells us that for large enough samples, these averages should follow a normal distribution, even if the original distribution from which they were chosen was not normal. Therefore we can use a normal distribution for our various tests.

15. **Consider the situation shown in Figure 6-9 where for the first time we are considering a diffusion situation in which mass is being injected at more than a single point. In particular, in this case equal masses M are injected at the two points shown, and we wish to analyze the flux passing by the flux measurement point shown on the figure.**

(a) **In a given time step, do you expect molecules to be passing by the flux measurement point in either direction?**

Yes, in both directions. In fact, each source should contribute molecules moving in both directions through this point, since no molecule from either source consistently moves always in the same direction.

(b) **What is the flux value q at the flux measurement point?**

By symmetry, it must be 0, since "neither side has any advantage over the other."

6.3 Modeling No-Flow Boundaries Using the Reflection Technique

1. **Consider a one-dimensional diffusion tube that is closed at one end. Suppose that a mass of 3 grams of material is injected instantaneously into the tube at a point 5 centimeters from the end and that the diffusion coefficient has the value 0.1 cm^2/sec. Determine the equation for the concentration profile as a function of location and elapsed time. In addition, through a sequence of graphs, illustrate the evolution of the shape of this profile over time. Pay particular attention to the evolution of any local maxima that may exist.**

The equation would be

$$C = \frac{3}{\sqrt{4\pi(.1)t}}e^{-\frac{(x-5)^2}{4(.1)t}} + \frac{3}{\sqrt{4\pi(.1)t}}e^{-\frac{(x+5)^2}{4(.1)t}}.$$

A sequence of graphs of this equation over the x-interval from 0 to 10 and for times 10, 50, 100, and 200 is shown in Figure 6-E. You can see that material accumulates near the closed end of the tube so that eventually diffusion off in the other direction causes the local maximum to migrate to the closed end, after which the concentration function is steadily decreasing away from that end.

2. **In the situation described by the previous problem, at what precise time would a local maximum concentration cease to exist anywhere within the tube?**

To solve this problem, we could find the location of the local maximum by considering the concentration function as a function of x with t as a parameter upon which the answer will depend. However, this approach is complicated by the fact that $x = 0$ is always a critical point (by symmetry), and so we need to track carefully the position of the local maximum between 0 and 5. An alternative approach is to note that the time in question is precisely the same as the time when the concentration at $x = 0$ has a local maximum as a function of time t. The reason for this is that when there is a local maximum for $x > 0$, there will still be a concentration gradient moving material towards the end, so the concentration at the end will be increasing. Afterwards, there will be a net movement away. Therefore the (local and absolute) maximum must occur just when the peak migrates to the end.

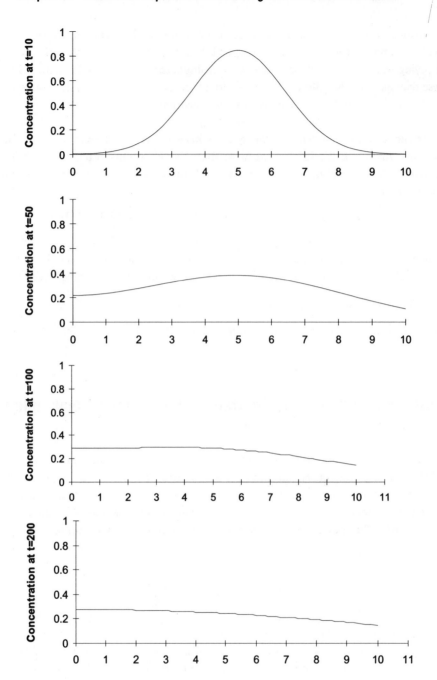

FIGURE 6-E
Evolution of concentration profile in one-ended tube (Exercise 1)

Therefore we look at the concentration at this one point, $x = 0$, as a function of t:

$$C = \frac{3}{\sqrt{4\pi(.1)t}}e^{-\frac{(0-5)^2}{4(.1)t}} + \frac{3}{\sqrt{4\pi(.1)t}}e^{-\frac{(0+5)^2}{4(.1)t}}$$

$$= \frac{2.676}{\sqrt{t}}e^{-\frac{62.5}{t}} + \frac{2.676}{\sqrt{t}}e^{-\frac{62.5}{t}}$$

$$= \frac{5.35}{\sqrt{t}}e^{-\frac{62.5}{t}}.$$

We now find that there is one critical value $t > 0$ where $C_t = 0$:

$$C_t = 5.35 \left(\frac{1}{\sqrt{t}} e^{-\frac{62.5}{t}} \cdot \frac{62.5}{t^2} - \frac{1}{2t^{3/2}} e^{-\frac{62.5}{t}} \right)$$

$$= 5.35 \frac{1}{\sqrt{t}} e^{-\frac{62.5}{t}} \left(\frac{62.5}{t^2} - \frac{1}{2t} \right)$$

$$0 = \frac{62.5}{t^2} - \frac{1}{2t}$$

$$t = 125,$$

and so it must be the maximum we are looking for.

3. Consider the two-dimensional diffusion layout shown in Figure 6-14. Find the concentration function for all points (x, y) for all points in time. You need not assume that the diffusion constants are the same in both the x- and the y-direction.

By superposition, this will just be

$$C = \frac{M}{4\pi t \sqrt{D_1 D_2}} e^{-\left(\frac{(x-3)^2}{4D_1 t} - \frac{(y-2)^2}{4D_2 t} \right)} + \frac{M}{4\pi t \sqrt{D_1 D_2}} e^{-\left(\frac{(x-2)^2}{4D_1 t} - \frac{(y+1)^2}{4D_2 t} \right)}$$

where the only difference between this and our original two-dimensional diffusion equation is the migration of the origin for each term so as to correspond to the source point.

4. Consider the following two-dimensional diffusion problem with a no-flow boundary. As usual, the x, y plane will describe the two-dimensional region available for diffusion, but assume in this case that an initial mass M is injected at the point (2,5), and that a physical boundary along the line corresponding to $y = -3$ causes this line to function as a no-flow boundary. Draw a figure for this situation and find the resulting solution function. Describe the region within which your solution function applies.

This situation is sketched in Figure 6-F. The solution applies to the unshaded region and is given by:

$$C = \frac{M}{4\pi t \sqrt{D_1 D_2}} e^{-\left(\frac{(x-2)^2}{4D_1 t} - \frac{(y-5)^2}{4D_2 t} \right)} + \frac{M}{4\pi t \sqrt{D_1 D_2}} e^{-\left(\frac{(x-2)^2}{4D_1 t} - \frac{(y+11)^2}{4D_2 t} \right)}.$$

5. Provide a clear explanation of how the principles described in this section are represented in the Gaussian plume equation.

Recall the Gaussian plume equation:

$$C = \frac{Q}{2\pi \sigma_y \sigma_z u} \left[e^{-\frac{y^2}{2\sigma_y^2}} \right] \left[e^{-\frac{(z-H)^2}{2\sigma_z^2}} + e^{-\frac{(z+H)^2}{2\sigma_z^2}} \right].$$

Considering Q/u as the source term (whose units are mass/length, so you can think of this as a source along the plume centerline consisting of a certain mass in each unit of length), we can rewrite the equation as:

$$C = \frac{Q/u}{2\pi \sigma_y \sigma_z} \left[e^{-\frac{y^2}{2\sigma_y^2} - \frac{(z-H)^2}{2\sigma_z^2}} \right] + \frac{Q/u}{2\pi \sigma_y \sigma_z} \left[e^{-\frac{y^2}{2\sigma_y^2} - \frac{(z+H)^2}{2\sigma_z^2}} \right]$$

which shows it is the superposition of two two-dimensional (x- and z-directions) diffusion solutions, one the real source along the plume centerline and one a virtual source along the reflection of the plume centerline down a distance H below the land surface.

6. Investigate the use of the term "reflection principle" for the technique described in this section. In particular, consider the one-dimensional diffusion situation of Figure 6-12 and assume that every time a

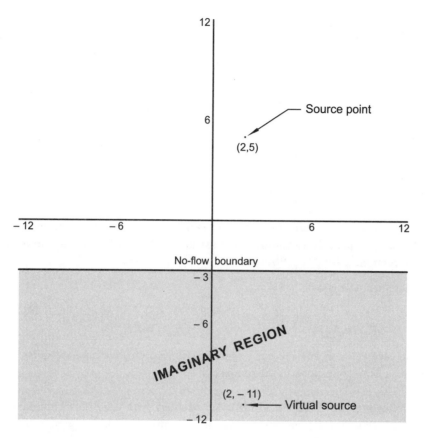

FIGURE 6-F
Situation described in Exercise 4

diffusing molecule wanted to move to the left through the boundary at the left end, it was reflected instead back towards the right. Show that the resulting solution actually equals the solution obtained in this section.

The whole concept of a virtual source was one whose corresponding flux through the boundary into the real, physical system would exactly equal the outgoing flux from the real system through the same boundary. Therefore the magnitude of this flux from the virtual source is identical to the effect of reflecting at the boundary any molecules that wanted to exit through it.

7.   [This exercise assumes that you completed the Exercise 23 on three-dimensional diffusion modeling in Section 6.1.] Imagine that a mass M of air pollutant is emitted into the atmosphere H meters above the ground. For simplicity in this particular problem, let us ignore all atmospheric effects (such air movement, stability classes, temperature, etc.), and let us consider this as a pure three-dimensional diffusion problem. Since this is not a continuous plume but only a momentary release, it is referred to as a "puff" release, and, in fact, you were asked to develop the general form of its solution in Section 6.1. However, in this case the ground behaves as a no-flow boundary. Taking this one additional factor into account, derive the appropriate solution to this problem.

By comparison with the cited exercise, we need first to move our source from the origin up to the point $(0, 0, H)$, and then all we need to add is a virtual source at a depth H below the ground:

$$C = \frac{M}{(4\pi t)^{3/2}\sqrt{D_1 D_2 D_3}}e^{\left(-\frac{x^2}{4D_1 t} - \frac{y^2}{4D_2 t} - \frac{(z-H)^2}{4D_3 t}\right)} + \frac{M}{(4\pi t)^{3/2}\sqrt{D_1 D_2 D_3}}e^{\left(-\frac{x^2}{4D_1 t} - \frac{y^2}{4D_2 t} - \frac{(z+H)^2}{4D_3 t}\right)}.$$

6.4 The Basic Partial Differential Equation for Diffusion Processes

1. Show that the one-dimensional diffusion equation is a solution to the one-dimensional diffusion PDE. (Hint: this computation may be somewhat complex so organize your work carefully to help you keep track of various terms.)

First we calculate C_t using the product rule for derivatives:

$$C_t = \frac{M}{\sqrt{4\pi Dt}} e^{-\frac{x^2}{4Dt}} \left(-\frac{x^2}{4D}\right)\left(\frac{-1}{t^2}\right) + e^{-\frac{x^2}{4Dt}} \frac{M}{\sqrt{4\pi D}}\left(-\frac{1}{2}t^{-\frac{3}{2}}\right)$$

$$= \frac{M}{\sqrt{4\pi Dt}} e^{-\frac{x^2}{4Dt}} \left[\frac{x^2}{4Dt^2} - \frac{1}{2t}\right]$$

$$= \frac{M}{\sqrt{4\pi Dt}} e^{-\frac{x^2}{4Dt}} \left[\frac{x^2 - 2Dt}{4Dt^2}\right].$$

Now proceeding to the x derivatives, we first have:

$$C_x = \frac{M}{\sqrt{4\pi Dt}} e^{-\frac{x^2}{4Dt}} \cdot \frac{-2x}{4Dt}$$

from which we get the second derivative by the product rule:

$$C_{xx} = \frac{M}{\sqrt{4\pi Dt}} \left[e^{-\frac{x^2}{4Dt}} \cdot \frac{-2}{4Dt} + \left(\frac{-2x}{4Dt}\right)e^{-\frac{x^2}{4Dt}} \cdot \frac{-2x}{4Dt}\right]$$

$$= \frac{M}{\sqrt{4\pi Dt}} e^{-\frac{x^2}{4Dt}} \left[\frac{-1}{2Dt} + \frac{4x^2}{(4Dt)^2}\right]$$

$$= \frac{M}{\sqrt{4\pi Dt}} e^{-\frac{x^2}{4Dt}} \left[\frac{-2Dt + x^2}{4D^2t^2}\right].$$

A visual comparison of C_t and C_{xx} makes it apparent that $C_t = DC_{xx}$, which is the differential equation we wanted to verify.

2. Can you find any other solutions to the one-dimensional diffusion PDE? Begin by looking for the simplest possible functions that work when substituted into the equation, and then gradually try to find some more complicated ones. Don't assume that they all have to have the very complicated form of our one-dimensional diffusion equation.

The following are all easily seen to be solutions:

$$x^2 + 2Dt, \qquad ax + b, \qquad e^{x+Dt}.$$

In addition, the derivatives of any solutions are also solutions (as long as the functions are sufficiently differentiable) since if we know that $C_t = DC_{xx}$, it follows that

$$C_{tx} = DC_{xxx}$$

$$(C_x)_t = D(C_x)_{xx}$$

and

$$C_{tt} = DC_{xxt}$$

$$(C_t)_t = D(C_t)_{xx}.$$

Therefore you might also expect (correctly) that integration (with appropriate constants of integration) could also generate additional solutions. Thus the solutions above along with our one-dimensional diffusion equation could all be used to generate many additional solutions.

3.   Suppose you have found a number of solutions to the one-dimensional diffusion PDE. Can you think of any way to modify or combine them in order to find yet more solutions?

Aside from the methods listed in the previous solution, any linear combination of solutions would also be a solution since the derivative operators are linear operators.

4.   [For readers with a background in linear algebra.] Considering the previous exercises, especially Exercise 3, can you describe in the language of linear algebra a key observation about the set of all solutions to the one-dimensional diffusion PDE?

This set is a vector space.

5.   [For readers with a background both in linear algebra and in differential equations.] Does your conclusion in Exercise 4, above, apply to the set of solutions to any differential equation, either ordinary or partial?

Definitely not. For example, if you have two solutions to a differential equation as simple as $\frac{dy}{dx} = 2$, their sum cannot be a solution since its derivative is 4.

6.   Explain why in the final sequence of steps just above, the limit expression actually does equal the indicated second derivative.

The quotient is the central difference formula for the derivative of C_x, which would yield C_{xx}. See Section 5.8 for further discussion of such formulas.

7.   Compare the derivation given above with the derivation presented in connection with Figure 6-3. Where do the derivations diverge from each other, and what are the precise objectives of the two different derivations?

There is considerable overlap, and the derivations diverge primarily in terms of their final steps. However, in the earlier section, the derivation did not make use of the diffusion principle, which had not yet been established for our specific one-dimensional diffusion equation. The point of the earlier derivation was to establish the connection between the derivative of flux and the time derivative of concentration: $-q_x = C_t$. In fact, we could have combined this with the diffusion principle to derive the one-dimensional diffusion PDE, as shown in the next solution. (But this was beyond our original needs, and by the time it came up again in the current section, it seemed better to provide a self-contained derivation.)

8.   The following equation was derived in Section 6.1, just prior to Exercise 15 of that section:

$$-\frac{\partial q}{\partial x} = \frac{\partial C}{\partial t}.$$

Show how the one-dimensional diffusion PDE could have been derived almost instantly from this equation.

$$-q_x = C_t$$
$$-(-DC_x)_x = C_t$$
$$DC_{xx} = C_t$$

9.   Describe precisely how in the derivation given in this section we have used the fundamental property that characterizes diffusion processes.

In performing our mass balance on the short section of the diffusion tube, we used the diffusion principle to represent the flux through each end of the section.

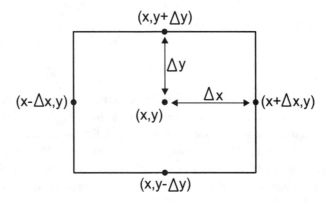

FIGURE 6-G

Spatial element for the mass balance in Exercise 10

10.   Draw a clear diagram and use it to derive the two-dimensional diffusion PDE for the case where the diffusion constant D is the same for all directions. Be sure to state the basic diffusion property precisely so that it is available at the key point in your derivation.

Figure 6-G shows an element for which a mass balance will be constructed. For two-dimensional problems, people often prefer to think of this element as a three-dimensional rectangular solid in 3-space, with the third dimension extending a distance of one unit perpendicular to the drawing. Thus the flux can indeed be thought of as the flow rate per unit cross-sectional area, similar to the approach taken in the much of Chapter 5 in connection with Laplace's equation. The concentration would also then be measured in mass per unit volume.

With the two diffusion directions being the x- and y-directions in the figure, the top and bottom faces of the solid element would exhibit no diffusion through them. Therefore, the mass balance would take the form:

rate of change of material within volume element

$= $ left face flow rate in $+$ right face flow rate in

$+$ bottom face flow rate in $+$ top face flow rate in

$$\frac{\partial}{\partial t}\big(C(x,y)\cdot 2\Delta x \cdot 2\Delta y \cdot 1\big) \approx q_1(x-\Delta x, y)(2\Delta y \cdot 1) - q_1(x+\Delta x, y)(2\Delta y \cdot 1)$$
$$+ q_2(x, y-\Delta y)(2\Delta x \cdot 1) - q_2(x, y+\Delta y)(2\Delta x \cdot 1)$$

$$\frac{\partial}{\partial t}\big(C(x,y)\cdot 2\Delta x \cdot 2\Delta y \cdot 1\big) \approx -DC_x(x-\Delta x, y)(2\Delta y \cdot 1) + DC_x(x+\Delta x, y)(2\Delta y \cdot 1)$$
$$- DC_y(x, y-\Delta y)(2\Delta x \cdot 1) + DC_y(x, y+\Delta y)(2\Delta x \cdot 1).$$

Here we have used the diffusion principle to write the fluxes in terms of the concentration gradients in each direction, and we have also suppressed the t variable in each function for simplicity because all calculations are at a single time t. The subscripts on the q's indicate the two diffusion directions. Now, dividing through by $2\Delta x \cdot 2\Delta y$ and taking the limit as these increments approach 0, we obtain

$$C_t = D(C_{xx} + C_{yy})$$

which is the two-dimensional diffusion equation.

11.   In the context of the previous problem, but assuming that different diffusion coefficients apply to the two diffusion directions, what should be the partial differential equation satisfied by the concentration?

The mass balance would now take the form

$$\frac{\partial}{\partial t}\big(C(x,y)\cdot 2\Delta x \cdot 2\Delta y \cdot 1\big) \approx -D_1 C_x(x-\Delta x, y)(2\Delta y \cdot 1) + D_1 C_x(x+\Delta x, y)(2\Delta y \cdot 1)$$
$$+ D_2 C_y(x, y-\Delta y)(2\Delta x \cdot 1) + D_2 C_y(x, y+\Delta y)(2\Delta x \cdot 1),$$

and so the limiting process would yield the equation

$$C_t = D_1 C_{xx} + D_2 C_{yy}.$$

12. Sometimes a multidimensional problem has sufficient symmetry to it so that it is "essentially" one-dimensional. For example, in a two-dimensional diffusion situation with the same diffusion constant for both directions and with a single initial mass M injected at one point, the concentration will obviously be the same all the way around every circle centered at the point of mass injection. Thus the only spatial variable that really should enter into the answer is the radius r. Considering C as a function of r in this symmetric case, draw a good diagram that enables you to derive directly from the mass balance the differential equation for C as a function of the single spatial variable r, as well as t.

See Figure 6-H. Here we perform the mass balance on the indicated element. As in Exercise 10, we should think of this sector as three-dimensional, with the third dimension, perpendicular to the drawing, having the value 1.

$$\frac{\partial}{\partial t}\big(C(r) \cdot r\theta \cdot 2\Delta r \cdot 1\big) \approx -q(r+\Delta r) \cdot (r+\Delta r)\,\theta \cdot 1 + q(r-\Delta r) \cdot (r-\Delta r)\,\theta \cdot 1$$

$$= DC_r(r+\Delta r) \cdot (r+\Delta r)\theta \cdot 1 - DC_r(r-\Delta r) \cdot (r-\Delta r)\theta \cdot 1$$

Now we divide through by $r\theta \cdot 2\Delta r$ and take the limit as $\Delta r \to 0$ to obtain

$$\lim_{\Delta r \to 0}\left[\frac{1}{r \cdot 2\Delta r}\big(DC_r(r+\Delta r) \cdot (r+\Delta r) - DC_r(r-\Delta r) \cdot (r-\Delta r)\big)\right]$$

$$= \lim_{\Delta r \to 0}\left[\frac{1}{r \cdot 2\Delta r}\big(DC_r(r+\Delta r) \cdot (r+\Delta r) - DC_r(r-\Delta r) \cdot (r+\Delta r) + DC_r(r-\Delta r) \cdot 2\Delta r\big)\right]$$

$$= \lim_{\Delta r \to 0}\left[\frac{r+\Delta r}{r} \cdot \frac{DC_r(r+\Delta r) - DC_r(r-\Delta r)}{2\Delta r} + \frac{1}{r} \cdot DC_r(r-\Delta r)\right]$$

$$= 1 \cdot DC_{rr} + \frac{1}{r} \cdot DC_r,$$

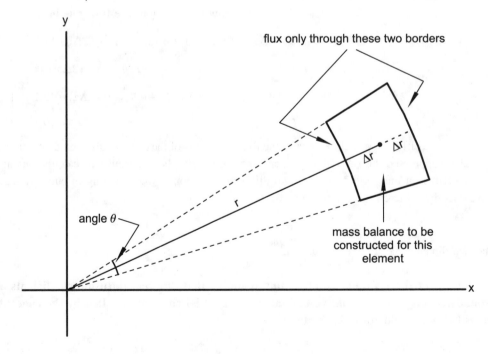

FIGURE 6-H
Problem with radial symmetry in Exercise 12

so that the final differential equation takes the form

$$C_t = D\left(C_{rr} + \frac{1}{r}C_r\right).$$

13. **As suggested in the previous exercise, the variables x and y are not the only variables with which you might describe the two-dimensional domain for diffusion. For example, since the polar coordinates r and θ can also be used to describe the two-dimensional domain, you would expect that the two-dimensional diffusion situation could be described in terms of these two new variables. In this exercise you are asked to develop the corresponding version of the two-dimensional diffusion PDE (that is, considering C as a function of r and θ) using two distinct lines of reasoning, namely:**

(a) by using the standard relations between the Cartesian coordinates x and y and the polar coordinates r and θ, in combination with the chain rule for partial derivatives, in order to transfer our original version of the two-dimensional diffusion PDE into these new coordinates;

To avoid ambiguity we need to use a different letter, say lower case $c(r, \theta)$ for the concentration function when considered as a function of a point's polar coordinates. Of course both C and c are also functions of time t, but we do not want to make the equations more complicated by writing that explicitly. All calculations are done at a single time t. Thinking of r and θ as functions themselves of x and y, given by the usual implicit relations

$$x = r\cos\theta$$
$$y = r\sin\theta,$$

we can write $C(x, y) = c\big(r(x, y), \theta(x, y)\big)$ and then substitute into the two-dimensional diffusion PDE, which applies to C. First we compute the derivatives:

$$C_x = c_r r_x + c_\theta \theta_x$$
$$C_{xx} = (c_r)_x r_x + c_r r_{xx} + (c_\theta)_x \theta_x + c_\theta \theta_{xx}$$
$$= (c_{rr}r_x + c_{r\theta}\theta_x)r_x + c_r r_{xx} + (c_{\theta r}r_x + c_{\theta\theta}\theta_x)\theta_x + c_\theta \theta_{xx}$$
$$= c_{rr}(r_x^2) + c_{r\theta}(2\theta_x r_x) + c_r(r_{xx}) + c_{\theta\theta}(\theta_x^2) + c_\theta(\theta_{xx})$$
$$C_{yy} = c_{rr}(r_y^2) + c_{r\theta}(2\theta_y r_y) + c_r(r_{yy}) + c_{\theta\theta}(\theta_y^2) + c_\theta(\theta_{yy}).$$

When the C_{xx} and C_{yy} terms are added, taking into account the specific coefficients obtained on the right from the polar coordinate relations (a standard partial differentiation exercise), considerable simplification takes place so that the result is:

$$C_{xx} + C_{yy} = c_{rr} + \frac{1}{r}c_r + \frac{1}{r^2}c_{\theta\theta}.$$

This yields the corresponding diffusion equation:

$$c_t = D\left(c_{rr} + \frac{1}{r}c_r + \frac{1}{r^2}c_{\theta\theta}\right)$$

which, once having been obtained, would often be written using the capital letter $C(r, \theta)$. Note that when the dependence on θ is not present, as in the previous exercise, this reduces to the equation obtained there.

(b) by beginning with a basic diagram of the physical situation in which you represent a spatial element whose geometry is appropriate to polar coordinates, and then you perform a mass balance on this element similar to the way it was done in the text for Cartesian coordinates.

This is very similar to the previous exercise except that the increment must be kept small in the θ direction as well. Thus the notation for the angles needs to be revised, as shown in Figure 6-I. For this case, there may be

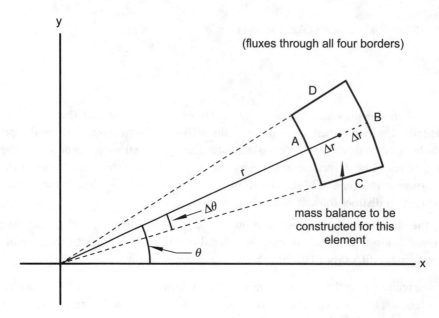

FIGURE 6-I

Typical mass balance element in polar coordinates (Exercise 13)

fluxes through all four borders of the element on the figure, and thus the mass balance takes the form:

rate of change of material within volume element

$$= \text{face } A \text{ flow rate in} + \text{face } B \text{ flow rate in}$$

$$+ \text{face } C \text{ flow rate in} + \text{face flow } D \text{ rate in}$$

$$\frac{\partial}{\partial t}(C(x,y) \cdot 2\Delta r \cdot 2r\Delta\theta \cdot 1) \approx q_r(r - \Delta r, \theta)(2(r - \Delta r)\Delta\theta \cdot 1) - q_r(r + \Delta r, \theta)(2(r + \Delta r)\Delta\theta \cdot 1)$$

$$+ q_\theta(r, \theta - \Delta\theta)(2\Delta r \cdot 1) - q_\theta(r, \theta + \Delta\theta)(2\Delta r \cdot 1)$$

$$\approx -DC_r(r - \Delta r, \theta)(2(r - \Delta r)\Delta\theta) + DC_r(r + \Delta r, \theta)(2(r + \Delta r)\Delta\theta)$$

$$- D\frac{C_\theta(r, \theta - \Delta\theta)}{r}(2\Delta r) + D\frac{C_\theta(r, \theta + \Delta\theta)}{r}(2\Delta r)$$

$$\approx -DC_r(r - \Delta r, \theta)(2(r - \Delta r)\Delta\theta) + DC_r(r + \Delta r, \theta)(2(r - \Delta r)\Delta\theta)$$

$$+ DC_r(r + \Delta r, \theta)(4\Delta r\Delta\theta) - D\frac{C_\theta(r, \theta - \Delta\theta)}{r}(2\Delta r) + D\frac{C_\theta(r, \theta + \Delta\theta)}{r}(2\Delta r).$$

In the last step, we have applied the diffusion principle, which for the r direction is basically the same as any other spatial direction. However, for the "θ" direction, where θ is an angle, not a distance, the diffusion principle would first need to be written, letting r be fixed, as:

$$q_\theta \text{ (meaning flux in the } \theta \text{ direction)} = -D\frac{dC}{d(r\theta)} = -D\frac{C_\theta}{r}.$$

Returning to the above sequence of equations, we now divide through by $4r\Delta r\Delta\theta$ and take limits, obtaining:

$$C_t = \lim_{\substack{\Delta r \to 0 \\ \Delta\theta \to 0}} \left[\frac{r - \Delta r}{r} \cdot \frac{DC_r(r + \Delta r, \theta) - DC_r(r - \Delta r, \theta)}{2\Delta r} \right.$$

$$\left. + \frac{DC_r(r + \Delta r, \theta)}{r} + \frac{DC_\theta(r, \theta + \Delta\theta) - DC_\theta(r, \theta - \Delta\theta)}{r^2 2\Delta\theta} \right]$$

$$= D\left(C_{rr} + \frac{1}{r}C_r + \frac{1}{r^2}C_{\theta\theta} \right).$$

14. **Beginning with a clear diagram, derive the three-dimensional diffusion PDE for the case where the diffusion constant D is the same in all directions.**

This is very similar to the two-dimensional case covered by Exercise 10 and is also similar to the corresponding three-dimensional derivation of Laplace's equation in Chapter 5. In fact, we will use the same figure as in Exercise 10, Figure 6-G, except that now we imagine that the element extends a distance $\pm \Delta z$ in the direction perpendicular to the drawing. The corresponding mass balance now must include two additional terms, for the two faces at $\pm \Delta z$, and the concentration must be written as a function of all three spatial variables:

rate of change of material within volume element

$$= \text{left face flow rate in} + \text{right face flow rate in}$$

$$+ \text{bottom face flow rate in} + \text{top face flow rate in}$$

$$+ \text{back face flow rate in} + \text{front face flow rate in}$$

$$\frac{\partial}{\partial t}\left(C(x,y,z) \cdot 2\Delta x \cdot 2\Delta y \cdot 2\Delta z\right) \approx q_1(x - \Delta x, y, z)(2\Delta y \cdot 2\Delta z) - q_1(x + \Delta x, y, z)(2\Delta y \cdot 2\Delta z)$$

$$+ q_2(x, y - \Delta y, z)(2\Delta x \cdot 2\Delta z) - q_2(x, y + \Delta y, z)(2\Delta x \cdot 2\Delta z)$$

$$+ q_3(x, y, z - \Delta z)(2\Delta x \cdot 2\Delta y) - q_3(x, y, z + \Delta z)(2\Delta x \cdot 2\Delta y)$$

$$\approx -DC_x(x - \Delta x, y, z)(2\Delta y \cdot 2\Delta z) + DC_x(x + \Delta x, y, z)(2\Delta y \cdot 2\Delta z)$$

$$- DC_y(x, y - \Delta y, z)(2\Delta x \cdot 2\Delta z) + DC_y(x, y + \Delta y, z)(2\Delta x \cdot 2\Delta z)$$

$$- DC_z(x, y, z - \Delta z)(2\Delta x \cdot 2\Delta y) + DC_z(x, y, z + \Delta z)(2\Delta x \cdot 2\Delta y).$$

In light of the similarity to several previous limits, it should be clear from the above that if we divide through by $8\Delta x \Delta y \Delta z$ and take the limit as the increments approach 0, we obtain the equation

$$C_t = D(C_{xx} + C_{xx} + C_{xx}),$$

and this is the desired result.

15. **State mathematically any and all boundary conditions associated with this basic problem.**

The initial concentration is 0 at every point $x \neq 0$, and a mass M is injected at time 0 at the point $x = 0$.

16. **[This is a very valuable but quite challenging problem. Some previous experience in solving differential equations, even if only ordinary differential equations, would be helpful.] It was verified in an earlier exercise that the one-dimensional diffusion equation is indeed a solution to the one-dimensional diffusion PDE. It has also been verified that this equation does satisfy the boundary condition described for our one-dimensional diffusion problem. However, it is one thing to be given a solution to a PDE and be asked simply to verify, by plugging it in, that it is a solution, and quite another thing to derive the solution in the first place. In this exercise you are asked to "forget" that you happen to know the solution to the one-dimensional diffusion problem, and you are asked to begin simply with the one-dimensional diffusion PDE and the corresponding boundary condition and show that that combination leads necessarily to a solution which turns out to be identical to the one-dimensional diffusion equation. (Hint: think about how the solution must depend on the units of the input values.)**

Let us begin by thinking of the same physical diffusion problem as worked out in two systems of units: the "U" units ($U = $ upper case) and the "L" units ($L = $ lower case). In the first set, the notation will be $C(X, T)$ for concentration at location X and time T. In these units the diffusion coefficient will be D. Similarly, in the L system of units, we will have a concentration $c(x, t)$ and a diffusion coefficient d. You may wish to think of the U units in terms of meters and minutes, and the L units as centimeters and seconds, at least for the first part of this discussion.

If one X-unit corresponds to a x-units, and one T-unit corresponds to b t-units, then we have the following conversion relationships:

$$x = aX$$

$$t = bT$$

$$d = \frac{a^2}{b}D, \quad \text{since diffusion coefficient units have form } \frac{\text{length}^2}{\text{time}}$$

$$c = \frac{1}{a}C, \quad \text{since concentration units are } \frac{\text{mass}}{\text{length}}$$

The concentration at a given physical location, all expressed in U units, can also be expressed as the concentration at the corresponding location in L units, and we get equality if we convert the latter back into U units. That is:

$$C(X, T) = a \cdot c(x, t) = a \cdot c(aX, bT).$$

This means that the actual numbers represented by each of these terms will be identical. Furthermore, if we choose our units carefully so that the quotient a^2/b turns out to be numerically equal to unity, then the values of D and d will be the same and so the unknown functions $C(X, T)$ and $c(x, t)$ will actually have exactly the same form, each of course in terms of its own variables. (For example, if we know that $C(X, T) = X^2 + Y^2$, it would follow that $c(x, y) = x^2 + y^2$.) This is because not only will the functions satisfy the same PDE, but also the same auxiliary conditions, since the relevant initial condition of a concentrated mass M at the origin does not depend on the units that we have been changing, nor does our implicit assumption of a 0 concentration at every point in the interior of the tube. This latter is a very important point because it is one of the two key places in which we are making use of our initial or boundary condition. This statement would not apply to other diffusion problem solutions resulting from any initial conditions depending on the distance scale.

The previous observation implies that $c(X, T) = C(X, T)$ so that $c(X, T) = a \cdot c(aX, bT)$. Remember that this is all under the special assumption about the relationship between and b (and one that is not satisfied by the units of meters, etc., that were mentioned above). However, looking at a and b in this last expression as parameters, and replacing a by $\sqrt{b}$, we have:

$$c(X, T) = \sqrt{b} \cdot c(\sqrt{b}X, bT).$$

Since this would hold for any values of X, T, and b, we can actually tailor our unit change to each point by letting $b = 1/T$, so that

$$c(X, T) = \frac{1}{\sqrt{T}} \cdot c\left(\frac{X}{\sqrt{T}}, 1\right)$$

where now the concentration function is written as a function of time and a new composite quotient variable. We could even multiply both sides through by the corresponding a, to change the c's to C's, so we would have U units throughout. But then, since the U units were arbitrary, we could have any units throughout and the statement would still be correct. With no further need to distinguish the U and L systems of units, we revert to our regular notation for the concentration function $C(x, t)$, which we now can write in the form:

$$C(x, t) = \frac{1}{\sqrt{t}} \cdot C\left(\frac{x}{\sqrt{t}}, 1\right) = \frac{1}{\sqrt{t}} \cdot G(z), \quad \text{where } z = \frac{x}{\sqrt{t}}.$$

We now apply the one-dimensional PDE to our expression for C in terms of $G(z)$, remembering to use the chain rule on the z terms:

$$0 = C_t - DC_{xx}$$

$$= \frac{1}{\sqrt{t}} \cdot G'(z) \cdot \left(\frac{-x}{2t^{3/2}}\right) - \frac{1}{2t^{3/2}} \cdot G(z) - \frac{D}{\sqrt{t}}\left(G''(z) \cdot \frac{1}{t}\right)$$

$$= \frac{-1}{2t^{3/2}} \cdot \left[\frac{x}{\sqrt{t}}G'(z) + G(z) + 2DG''(z)\right].$$

Therefore, we obtain the ordinary differential equation

$$0 = 2DG''(z) + zG'(z) + G(z).$$

We want to solve this second-order linear homogeneous ordinary differential equation for $G(z)$. (By the way, from basic ordinary differential equation theory, this equation is easily seen to have a unique solution for initial conditions on $G(0)$ and $G'(0)$, the kinds of conditions we will apply. The set of all solutions is a two-dimensional vector space.)

There are various approaches to such equations, but this one will easily yield to a power series approach, which is one of the most straightforward methods. Rather than plug in a power series and then match up terms (which would also work fine), we will actually find directly the particular solution that applies to our particular problem. In particular, the Taylor series for G around $z = 0$ has the general form

$$G(z) = G(0) + G'(0)z + \frac{G''(0)}{2!}z^2 + \frac{G'''(0)}{3!}z^3 + \frac{G^{(4)}(0)}{4!}z^4 + \cdots,$$

and therefore we need to find the successive derivatives of G at 0. We do this by rewriting the differential equation for G so as to express the second derivative in terms of lower-order derivatives, and then we repeatedly differentiate and evaluate each of the resulting derivatives at 0:

$$G'' = \frac{-1}{2D}(zG' + G) \qquad G''(0) = \frac{-1}{2D}(G(0))$$

$$G''' = \frac{-1}{2D}(zG'' + 2G') \qquad G'''(0) = \frac{-1}{2D}(2G'(0))$$

$$G^{(4)} = \frac{-1}{2D}(zG''' + 3G'') \qquad G^{(4)}(0) = \frac{-1}{2D}(3G''(0))$$

from which the pattern is clear. Therefore, all we need are the two starting points $G(0)$ and $G'(0)$. Recall that by definition $G(z) = C(z, 1)$, and so $G(0)$ is just the unknown non-zero constant $C(0, 1)$, which we shall call A. The derivative $G'(0)$ is just

$$G'(0) = \frac{d}{dz}C(z, 1)\bigg|_{z=0} = C_x(0, 1) = 0$$

by the symmetry of the concentration, as a function of x, at 0. Thus all the odd power coefficients in the Taylor series are zero, and the series becomes:

$$G(z) = A - A\frac{1}{2D \cdot 2!}z^2 + A\frac{3 \cdot 1}{(2D)^2 4!}z^4 - A\frac{5 \cdot 3 \cdot 1}{(2D)^3 6!}z^6 + \cdots$$

$$= A\left[1 - \frac{z^2}{2D \cdot 2} + \frac{1}{(2D)^2 \cdot 2^2 \cdot 2!}z^4 - \frac{1}{(2D)^3 \cdot 2^3 \cdot 3!}z^6 + \cdots\right]$$

$$= A\left[1 - \left(\frac{z}{2\sqrt{D}}\right)^2 + \frac{1}{2!}\left(\frac{z}{2\sqrt{D}}\right)^4 - \frac{1}{3!}\left(\frac{z}{2\sqrt{D}}\right)^6 + \cdots\right]$$

$$= A\left[1 + \left\{-\left(\frac{z}{2\sqrt{D}}\right)^2\right\} + \frac{1}{2!}\left\{-\left(\frac{z}{2\sqrt{D}}\right)^2\right\}^2 + \frac{1}{3!}\left\{-\left(\frac{z}{2\sqrt{D}}\right)^2\right\}^3 + \cdots\right]$$

$$= Ae^{-\left(\frac{z}{2\sqrt{D}}\right)^2}$$

$$= Ae^{-\frac{z^2}{4D}}$$

$$= Ae^{-\frac{x^2}{4Dt}}.$$

where we have recognized the usual series for the exponential function once the complexities of its argument in this case were all grouped together.

Using this expression in our equation for $C(x,t)$, we now have:

$$C(x,t) = \frac{1}{\sqrt{t}} \cdot G(z) = \frac{1}{\sqrt{t}} \cdot Ae^{-\frac{x^2}{4Dt}}.$$

For any fixed t, we will use the fact that the integral of this expression from minus infinity to plus infinity must be the total mass M in the system. A particularly convenient t to work at would be one that makes the integral a well-known integral, like the standard normal distribution. Therefore, taking $t = 1/2D$, we have

$$M = \int_{-\infty}^{\infty} C(x,t)\, dx = \int_{-\infty}^{\infty} \sqrt{2D} \cdot Ae^{-\frac{x^2}{2}}\, dx = \sqrt{2D} \cdot A \cdot \int_{-\infty}^{\infty} e^{-\frac{x^2}{2}}\, dx = \sqrt{2D} \cdot A \cdot \sqrt{2\pi}$$

so that we find A to be given by

$$A = \frac{M}{\sqrt{4\pi D}}$$

and hence, finally (!),

$$C(x,t) = \frac{M}{\sqrt{4\pi Dt}} e^{-\frac{x^2}{2}}.$$

17. Consider a one-dimensional diffusion situation for a "semi-infinite" tube, meaning one that is infinite to the right but has a closed left end. Suppose that by some ingenious apparatus, we can always maintain the concentration of diffusing material at the constant value C_0 just at the left end point. Furthermore, let us suppose that we begin the experiment with a concentration 0 everywhere else. What would happen is that material would gradually flow from the left end point towards the right through the tube and we would have to keep replenishing it at the left end point to keep the concentration in the immediate "infinitesimal" neighborhood of that point at C_0. So we would constantly be injecting mass, and mass would continue to be moving from left to right throughout the tube. Find the concentration function $C(x,t)$ and the rate at which we would have to be supplying mass to maintain this situation.

If you consider the unknown function $M(s)$ to represent the rate at which mass is being supplied at the left end at time s, then you can think of this problem as the limiting case of the superposition of solutions of individual problems, each involving a short time interval Δs, during which an amount of mass $M(s)\Delta s$ is released. But this latter is one of our typical one-dimensional diffusion problems, whose solution at some time t is

$$C = \frac{2M(s)\Delta s}{\sqrt{4\pi D(t-s)}} e^{-\frac{x^2}{4D(t-s)}}.$$

(The factor 2 in the numerator comes from the reflection principle, since the tube is closed at one end.) To find the complete solution, we need to add up all solutions that correspond to s values that could influence the result at time t, namely, $s \le t$. This limit is simply the integral:

$$C(x,t) = \int_0^t \frac{2M(s)}{\sqrt{4\pi D(t-s)}} e^{-\frac{x^2}{4D(t-s)}}\, ds.$$

This integral cannot be determined analytically but represents a perfectly acceptable form of solution. The only problem is that we need to determine the right function $M(s)$ so as to maintain the correct constant concentration value C_0 at the left endpoint. This constraint takes the form:

$$C_0 = C(0,t) = \int_0^t \frac{2M(s)}{\sqrt{4\pi D(t-s)}} e^{-\frac{0^2}{4D(t-s)}}\, ds$$

$$= \frac{1}{\sqrt{\pi D}} \int_0^t \frac{M(s)}{\sqrt{t-s}}\, ds.$$

Focusing solely on the integral, we can determine $M(s)$ by changing the variable of integration to $u = s/t$, as follows:

$$\int_0^t \frac{M(s)}{\sqrt{t-s}} ds = \int_0^1 \frac{M(ut)}{\sqrt{t-ut}} d(ut) = \int_0^1 \frac{M(ut)\sqrt{t}}{\sqrt{1-u}} du$$

so that it is at least obvious that any function of the form $M(ut) = A/\sqrt{ut}$ reduces the integral to a constant (i.e., independent of t).

Could there be any others too? No. To make the integral a constant, we certainly need the numerator to be function of u alone, say, $f(u)$. But then it follows that

$$M(ut) \cdot \sqrt{t} = f(u)$$

$$M(t) = f(1) \cdot \frac{1}{\sqrt{t}} = A \cdot \frac{1}{\sqrt{t}}, \quad \text{for } A \equiv f(1)$$

$$M(ut) = A \cdot \frac{1}{\sqrt{ut}}.$$

Thus, with our mass supply function having this form, we return to the previous sequence of equations in order to determine the correct constant A:

$$C_0 = \frac{A}{\sqrt{\pi D}} \int_0^1 \frac{1}{\sqrt{u(1-u)}} du$$

$$= \frac{A}{\sqrt{\pi D}} \left[\arcsin(2u - 1) \right]\big|_0^1, \quad \text{(using a trigonometric substitution)}$$

$$= \frac{A}{\sqrt{\pi D}} \cdot \pi$$

$$A = \frac{C_0\sqrt{D}}{\sqrt{\pi}}.$$

Therefore, in our original terminology, the concentration at the left end can be maintained at the constant level C_0 by supplying mass at the rate $M(t)$ given by

$$M(t) = \frac{C_0\sqrt{D}}{\sqrt{\pi t}}.$$

You can see that the supply rate decreases with time as the material "backs up" a bit in the tube, causing it to be removed more slowly from the end where it is being supplied. The corresponding concentration throughout the tube would then be given by

$$C(x,t) = \int_0^t \frac{2\dfrac{C_0\sqrt{D}}{\sqrt{\pi s}}}{\sqrt{4\pi D(t-s)}} e^{-\frac{x^2}{4D(t-s)}} ds$$

$$= \frac{C_0}{\pi} \int_0^t \frac{1}{\sqrt{s(t-s)}} e^{-\frac{x^2}{4D(t-s)}} ds.$$

There are many ways to transform this result that we will not pursue, except to give a typical result in the form of an integral with respect to x:

$$C(x,t) = C_0 - 2C_0 \int_0^x \frac{1}{\sqrt{4\pi Dt}} e^{-\frac{x^2}{4Dt}} dx.$$

This form actually can also be derived directly by pursuing the idea, raised earlier in the solution to Exercise 2, that the integration of solutions to the diffusion PDE leads to other solutions. (Then one needs to modify the results slightly to achieve the desired boundary conditions.) In any case, these integrals can all also be transformed to the variable of integration

$$z = \frac{x}{\sqrt{2Dt}}$$

and then written in terms of the classical error function, erf z, which is their most common final form.

See Section 7.3.2 for an alternative approach to this kind of problem in the framework of its heat flow analog.

18.   Consider the original infinitely long one-dimensional diffusion tube where the initial mass M, rather than being injected at the point $x = 0$, is initially distributed uniformly over the interval from $x = -1$ to $x = 1$. (That is, the initial concentration in this particular interval of the tube would be $M/2$.) Assuming that the initial concentration is 0 elsewhere, find the concentration function $C(x, t)$ that would describe the concentration function at all times $t > 0$.

We will treat this as the superposition of the solutions of individual problems of the form we have generally been treating. In particular, divide the source interval into n equal subintervals, each of length $\Delta x = 2/n$. The amount of initial mass over each subinterval will constitute the source for one of our more typical problems, which you can thus think of as a source mass M/n essentially concentrated at some point in the subinterval.

In general terms, the solution at a given location u for such a source at x would be given by our one-dimensional diffusion equation:

$$C = \frac{\left(\dfrac{M\Delta x}{2}\right)}{\sqrt{4\pi Dt}} e^{-\frac{(u-x)^2}{4Dt}},$$

and so the limit of the sum of such solutions would be the corresponding integral as x goes from -1 to $+1$:

$$C(u, t) = \int_{-1}^{1} \frac{M/2}{\sqrt{4\pi Dt}} e^{-\frac{(u-x)^2}{4Dt}} \, dx.$$

Although the integral cannot be determined analytically, it can be calculated numerically or even from tables of the normal distribution or the related "error function." This integral would be a reasonable answer to the problem, but let us investigate it further by changing the variable of integration to $y = u - x$:

$$C(u, t) = \int_{u-1}^{u+1} \frac{M/2}{\sqrt{4\pi Dt}} e^{-\frac{y^2}{4Dt}} \, dy = \frac{1}{2} \int_{u-1}^{u+1} \frac{M}{\sqrt{4\pi Dt}} e^{-\frac{y^2}{4Dt}} \, dy,$$

which shows it to be the same as the average value of the concentration we would have found in the interval from $u - 1$ to $u + 1$ from an original source M concentrated at the origin. So the concentration at a point resulting from a distributed (i.e., spread out) source is similar to the average concentration over an interval resulting from a concentrated source.

19.   Describe a physical ground-water problem that has the following two characteristics: a) the hydraulic head values are a function not only of spatial location but also of time, and b) the hydraulic head profile actually does approach a steady state value for large values of t.

When you start pumping a well, the cone of depression does not form suddenly because the system is "buffered" against instant change by the water stored within it. As has been mentioned earlier, this is true even for a confined aquifer where storage capacity is not provided by a fluctuating water table. The elasticity of the geologic materials is the main source of this local buffering effect, although water is not perfectly incompressible and thus also contributes to some effect. As long as the well pumps steadily, an equilibrium condition will generally be approached where the flow rate is balanced against the other boundary conditions on the system.

20.   With reference to the previous exercise, give an example of a real, nontrivial physical ground-water situation that satisfies condition a in the exercise but not condition b.

The first example generally thought of is a well whose pumping rate varies with time, but that would be the obvious "trivial" example. An important real example is the overpumping of an aquifer, such as is happening in the plains states with the famous Ogallala Aquifer, which underlies much of the Midwest. Water withdrawals exceed the capability of the recharge areas for the aquifer, and thus the water levels just keep falling. This is a very serious problem in other areas as well and is sometimes called the "mining" of ground water. In principle, of course, an equilibrium should eventually be reached, but this is such a slow process and the system is still so far from it, that for all practical purposes it is just a constantly changing system, even if the water withdrawal rates were to stay relatively constant for a long time.

6.5 Guide to Further Information

(No exercises.)

7

Additional Topics in Hazardous Materials Modeling

7.1 Discharge Submodels

1. For the situation shown in Case A of Figure 7-3, determine the final velocity of the mass as it reaches the base level. Express your answer in terms of the height h and the gravitational constant g. (No time values should show up in your final answer.) Describe the relationship between this answer and the equation of Torricelli given above for a somewhat different situation.

A formula often encountered in elementary physics is $v = \sqrt{2as}$, which in the present notation takes the form $v = \sqrt{2gh}$. Perhaps the easiest derivation is by energy considerations. At the top, the mass has no kinetic energy, but it does have potential energy (with respect to the base level) of mgh. On the contrary, at the bottom it has no potential energy left, but its kinetic energy is now $\frac{1}{2}mv^2$. By the conservation of energy, we have:

$$\frac{1}{2}mv^2 = mgh$$
$$v = \sqrt{2gh}.$$

2. Consider the situation shown in Case B of Figure 7-3. Here a whole stack of equal masses are allowed to fall freely, beginning from rest. What will be the velocity of the topmost mass at the time it reaches the base level?

It will be given by $v = \sqrt{2gh}$, the same as in the previous exercise. The masses below it have no effect. Like an elevator dropping in free fall, the lower mass no longer exerts a force on the one above it, which drops independently.

3. If you tie a weight onto a piece of rope and swing it around over your head, the weight is constantly accelerating because its velocity is changing. (The velocity is not changing in magnitude, but it is changing in direction.) This is, of course, consistent with Newton's Law, $F = ma$, for there is a real force that you can feel your hand and arm exerting as you swing the weight around. Calculate the amount of work you are performing on the weight. You may assume that the weight has mass m, the rope is 3 feet long, and the weight is rotating at one revolution per second.

In the idealized situation, you are performing no work on the mass, as work is force times distance, and your force is not moving through any distance. (The motion is always perpendicular to the force.) On the other hand, in order to keep the weight swinging around, you do have to keep applying some additional effort to cover the frictional losses. If you were to watch your hand carefully, you would be moving it slightly to keep the weight going. This movement would not be perpendicular to the weight's movement, and would constitute work.

4.   Clarify the logic in the previous paragraph concerning the issue of whether the slide performs any work on the mass. In particular, formulate a precise mathematical expression for work and use it to explain your argument clearly. Relate this situation to that given in Exercise 3.

Let θ be the (possibly changing) angle between a force and the actual direction of movement of a mass. Then the work performed would be $F \cos \theta$ times the distance. More generally, for a changing path, one would need to take the line integral of the dot product of the force vector and the displacement vector along the trajectory C. That is,

$$W = \int_C F \cdot ds = \int_C F \cos \theta \, ds.$$

Since θ would always be $90°$ in the given situation, this integral would be 0.

5.   In Case C, the mass is actually rotated as it slides down. Where does this fit into the application of the conservation of energy principle? What reasonable assumptions could you make to avoid this issue?

It was ignored in the idealized situation discussed in the text. However, to be precise, there is an angular acceleration of the mass from its stationary initial condition, and thus there is a growing angular component to its kinetic energy. Since this too must come from the initial potential energy, the amount of energy actually converted into linear movement will be slightly less than was calculated above, making the velocity at the bottom slightly less than $\sqrt{2gh}$. A reasonable assumption that would cause this issue to disappear is that the mass is concentrated at a point.

6.   Torricelli's equation applies to the general situation shown in Figure 7-2. Explain the key physical differences in fluid flow between this situation and that shown in Figure 7-3, Case F. Can you work through these differences, either heuristically or mathematically, to develop a convincing argument of why Torricelli's equation is a plausible or reasonable answer to the original problem?

Figure 7-3, Case F, depicts a relatively narrow tube within which flow conditions are generally uniform over a cross section. The tank in Figure 7-2 seems much more complex, and flow conditions might be expected to be quite different from point to point. Nevertheless, assuming relatively smooth and non-turbulent flow, if you were to sketch in the flow lines in this figure, they might have the shape suggested in Figure 7-7, later in Volume 1. Now restrict your attention to an individual flow line and then to a narrow 'flow tube' centered on that flow line. This is similar to the tube shown in Figure 7-3; and as long as the fluid can move along it with negligible friction effects from the adjacent fluid, the velocity at the bottom should be $\sqrt{2gh}$. Since we could repeat this for the tube around any flow line, always getting the same answer, the velocity at the discharge point would be relatively uniform with the value $\sqrt{2gh}$. (See the ensuing text for further elaboration of the flow tube concept.)

7.   What would Torricelli's equation predict as the discharge velocity from a half-inch diameter hole in the side of a vertical cylindrical tank of water, given that the water level in the tank is 25 feet high and the hole in the side is 3 feet high? What would be the effect on discharge velocity of doubling the diameter of the hole?

The hole diameter does not enter the calculation and is thus irrelevant (within the limits of the model). In this case $h = 22$ ft, and so we have

$$v = \sqrt{2gh} = \sqrt{2 \times 32 \text{ ft/sec}^2 \times 22 \text{ ft}} = 37.52 \text{ft/sec.}$$

8.   How would you expect Torricelli's equation to fit into the determination of actual volumetric or mass discharge rates from holes in tanks?

One would expect to take the discharge velocity and multiply by the area of the opening to get the total discharge rate (in units of volume per time). If you then multiplied by the density, you should get units of mass per time.

Same height at equilibrium

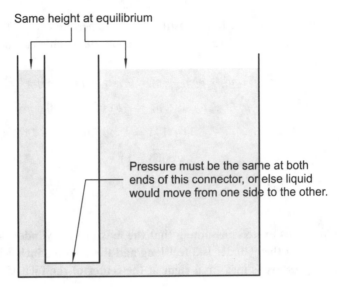

Pressure must be the same at both ends of this connector, or else liquid would move from one side to the other.

FIGURE 7-A
Equilibrium condition between narrow and wide columns of liquid (Exercise 9)

However, despite this expectation, it is not quite right, as the momentum of the flow at the edges of the opening cause it to have an "effective" area less than its total physical area. This decreases the flow rate proportionately. (In addition, there is a slight frictional effect at the opening.) This is discussed later in this section.

9. **Consider two vertical tanks of water, each 40 feet high and each filled to the top. Tank A has a diameter of 80 feet. Tank B has a diameter of 20 feet. Let P_A and P_B represent the respective pressures inside the two tanks right at a point where the sides and bottoms meet. Determine the relation between P_A and P_B.**

They are equal. To see this, consider Figure 7-A, where you know by experience that the water level will be the same in both columns when they are in equilibrium. The only way this can happen is if there is a balance of pressures at the ends of the small tube connecting the them, meaning that the pressures are the same at the bottom of the narrow and the wide columns.

10. **Consider the situation represented in Figure 7-6, which shows a large-diameter cylinder through which a piston is moving to squeeze oil out through a small nozzle at the right end. The force on the piston is such that it maintains a constant pressure of 60 psi (pounds per square inch) within the large cylinder, and the diameter of the cylinder is so much larger than the nozzle that the velocity of the piston (or the fluid in the large cylinder) can be treated as negligible. At the tip of the nozzle, the pressure may be assumed to be atmospheric (14.7 psi). Find the velocity as the oil exits the nozzle. (Hint: consider the cylinder/nozzle combination as a flow tube and study the streamline right along the center using Bernoulli's equation. Assume that the oil weighs 50 lb/ft^3. If your answer turns out less than 20 ft/sec, then you have made a mistake somewhere.)**

An easy error is to use pound units for both force and mass. To avoid this, convert the density of the oil to slugs per cubic foot:

$$\text{force (pounds)} = \text{mass (slugs)} \times g \ (\text{ft/sec}^2)$$

$$50 = m \times 32$$

$$m = 1.56 \text{ slugs}.$$

Thus the density of the oil would be 1.56 slugs/ft^3. Also, you need to convert the pressure units to force per square foot by multiplying by 144 in^2/ft^2. Therefore, we now apply Bernoulli's equation:

$$\tfrac{1}{2}\rho v_L^2 + \rho g y_L + p_L = \tfrac{1}{2}\rho v_R^2 + \rho g y_R + p_R$$

$$\tfrac{1}{2}\rho \cdot 0^2 + \rho g \cdot 0 + p_L = \tfrac{1}{2}\rho \cdot v_R^2 + \rho g \cdot 0 + p_R$$

$$60 \times 144 = \tfrac{1}{2} \times 1.56 \times v_R^2 + 14.7 \times 144$$

$$\sqrt{8363} = v_R^2$$

$$91.4 \text{ ft/sec} = v_R.$$

11.   Repeat the previous exercise assuming that the axis of the cylinder is now pointing up to the right at an angle of 45° and that the cylinder is 3 feet long and the nozzle 3 inches long. (Assume, to be precise, that the 60-psi constant pressure is measured right at the center of the face of the piston.)

In this case, the difference in heights keeps at least one of the potential energy terms from canceling out. Taking the base level as the center of the piston, the height of the tip of the nozzle would be

$$y_R = \left(3 + \frac{3}{12}\right)\cos 45° = \frac{13}{4} \cdot \frac{\sqrt{2}}{2} = 2.30 \text{ ft.}$$

The application of Bernoulli's equation now proceeds as follows:

$$\tfrac{1}{2}\rho v_L^2 + \rho g y_L + p_L = \tfrac{1}{2}\rho v_R^2 + \rho g y_R + p_R$$

$$\tfrac{1}{2}\rho \cdot 0^2 + \rho g \cdot 0 + 60 \times 144 = \tfrac{1}{2} \cdot 1.56 \cdot v_R^2 + 1.56 \cdot 32 \cdot 2.30 + 14.7 \times 144$$

$$\sqrt{8248} = v_R^2$$

$$90.8 \text{ ft/sec} = v_R.$$

Thus one sees a slight slowing of the nozzle speed from the upward orientation of the apparatus, as would be expected.

12.   Suppose that you were going to be working a number of problems that were expressed in the same original units as the previous two exercises, or that you were programming a computer model for which you would like to allow the user to input these units. That is, both force and mass would be given in pounds, and pressure would be given in psi. Distance units would still be in feet and time in seconds. Modify Bernoulli's equation so that it applies directly to this situation. (Keep your result handy for future reference.)

Following the logic of the previous problems and incorporating obvious simplifications, we would have:

$$\frac{\rho}{64}v_2^2 + \rho y_2 + 144 p_2 = \frac{\rho}{64}v_1^2 + \rho y_1 + 144 p_1$$

13.   Consider the situation shown in Figure 7-7 except for the following modification. The vent on the tank has been sealed, and, instead, the contents are kept under a constant pressure p_b. Find an expression for the discharge velocity. (Note: the contents might be kept under pressure either by the vapor pressure of the material itself, especially if it is a material with a low boiling point, or by an externally supplied pressurized gas source. The latter is generally accomplished with an inert gas like nitrogen so as to keep oxygen out of the space above the liquid and thus minimize fire and explosion potential.)

Making the appropriate modifications to the derivation in the text:

$$\tfrac{1}{2}\rho 0^2 + \rho g h + p_b = \tfrac{1}{2}\rho v^2 + \rho g 0 + p_a$$

$$\rho g h + (p_b - p_a) = \tfrac{1}{2}\rho v^2$$

$$2gh + \frac{2(p_b - p_a)}{\rho} = v^2$$

$$\sqrt{\frac{2gh + 2(p_b - p_a)}{\rho}} = v.$$

14.   Ammonia is a compound that has a low boiling point (–28°F), so that at ambient temperatures it will produce vapor pressures well in excess of atmospheric pressure. In particular, at 80°F its vapor pressure is about 150 psi, or about ten times atmospheric pressure. Suppose you have a vertical tank of ammonia that is 6 feet in diameter and 20 feet high, and that the internal pressure is its vapor pressure at the ambient temperature of 80°F. Assume that it is half full. If a small gauge fitting at the bottom of the tank breaks off due to undetected corrosion, what would our model predict the velocity of discharge to be? The specific gravity of ammonia is about 0.68.

We could use the results from either of the previous two problems, but for here we will work in the framework of Exercise 12, which has the unit conversions incorporated. Note that ammonia must weigh 68% as much as water, and hence $0.68 \times 62.4 = 42.4$ lb/ft^3. Thus we have

$$\frac{\rho}{64}v_2^2 + \rho y_2 + 144 p_2 = \frac{\rho}{64}v_1^2 + \rho y_1 + 144 p_1$$

$$\frac{42.4}{64} \cdot 0^2 + 42.4 \cdot 10 + 144 \cdot 150 = \frac{42.4}{64}v_1^2 + 42.4 \cdot 0 + 144 \cdot 14.7$$

$$30049 = v_1^2$$

$$173.3 \text{ ft/sec} = v_1.$$

We do not have the methodology available to evaluate whether our model, based on Bernoulli's equation, will really apply very well to this real case, especially when the predicted discharge velocity is so high. However, one would expect that if there were other factors that disrupted the flow at these speeds, they would only slow it down, not speed it up. Therefore, this discharge velocity would still serve as a conservative estimate of the real velocity.

15.   Repeat the previous exercise except for the following single change: the tank is now a horizontal tank of the same dimensions. (This would be more common for an ammonia tank.)

The only change in this case is the height factor, which is 3 feet since the tank is half full:

$$\frac{\rho}{64}v_2^2 + \rho y_2 + 144 p_2 = \frac{\rho}{64}v_1^2 + \rho y_1 + 144 p_1$$

$$\frac{42.4}{64} \cdot 0^2 + 42.4 \cdot 3 + 144 \cdot 150 = \frac{42.4}{64}v_1^2 + 42.4 \cdot 0 + 144 \cdot 14.7$$

$$29601 = v_1^2$$

$$172.0 \text{ft/sec} = v_1.$$

16.   Torricelli's equation may be thought of as a special case of your solution to Exercise 13, above, when the pressure forces cancel out and do not affect the discharge rate, so that the gravitational force is the driving force for discharge. At the other end of the spectrum is the case when the pressures inside and outside the tank are so different that they far overshadow the effects of gravity. In this case, how would your solution to Exercise 13 simplify?

The height factor would drop out and the result would be:

$$\sqrt{\frac{2(p_b - p_a)}{\rho}} = v.$$

17. Apply your solution to Exercise 16 to the problem given in Exercise 14, and relate the result to your answers to both Exercises 14 and 15.

We apply Exercise 16 to obtain:

$$v = \sqrt{\frac{2(p_b - p_a)}{\rho}} = \sqrt{\frac{2(150 - 14.7)(144)}{\frac{(.68)(62.4)}{32}}} = 171.4\text{ft/sec}.$$

This is slightly less than in Exercises 15 and 16 because the gravitational effect has been dropped, but the dominant factor has been incorporated in the same way in all three problems.

18. A number of assumptions were made in the derivation of Bernoulli's equation. These involved both the physical basis for the problem as well as the mathematical framework for analyzing it. For each of the assumptions listed below, clearly identify how it fits into the derivation:

a) Incompressibility of the fluid.

If this were not to hold, then the original mass conservation equation would not simplify to $A_1 v_1 = A_2 v_2$, and we would not have the basic equation of continuity. This latter was applied to the approximations to the integrals.

b) Zero viscosity of the fluid.

Frictional losses in a viscous fluid would render the energy balance in our derivation inaccurate. Such losses would need to be incorporated on the right-hand side.

c) Small (eventually "infinitesimal") cross section of the flow tube.

This allowed the integrands in the energy balance to contain a single pressure value p(s) applicable to each cross-section.

d) Small (eventually "infinitesimal") length of the fluid element.

This allowed the integrals to be replaced by the indicated products (e.g., using the mean value theorem for integrals, as discussed in the next exercise).

19. [Involves theoretical aspects of calculus.] In the derivation of Bernoulli's equation, one step involved the approximation

$$\int_a^b p(s)A(s)\,ds \cong p_1 A_1 (b - a)$$

as well as another similar approximation. Although it is convenient to speak in terms of such approximations, it is good to know how to test them rigorously by more precise mathematical analysis. Use the mean value theorem for integrals to simplify the expression on the left, and then show how the result would fit into the subsequent part of the derivation of Bernoulli's equation.

The mean value theorem for integrals says that for a continuous function $f(x)$ on a closed interval, there is some point c within the interval such that

$$\int_a^b f(x)\,dx = f(c) \cdot (b - a).$$

Assuming on physical grounds that the functions $p(s)$ and $A(s)$ are continuous, we apply the theorem to their product:

$$\int_a^b p(s)A(s)\,ds = p(c)A(c)(b-a), \quad \text{for some } c \text{ between } a \text{ and } b$$

$$\cong p_1 A_1 (b-a).$$

If the interval is short, the values of pA at the two points will be close. Furthermore, by continuity, $p(c)A(c)$ approaches $p_1 A_1$ as the interval approaches 0 in length.

20. Give an intuitive explanation of why the density of the liquid does not enter into Torricelli's equation. Wouldn't you expect that a heavier liquid would exert more pressure and hence flow through the discharge opening faster?

There is an equal but opposite counterbalancing feature, namely, the fact that a denser liquid would have more mass per unit volume that would need to be accelerated. Therefore both the force and the inertia would be larger, and the derivation shows that they balance out. For the same reason, *as long as you ignore frictional effects,* a feather and a cannonball dropped from the top of a building should fall at the same rate. (But frictional effects, as in this last example, become a more dominant factor when the mass is small.)

21. If you were to calculate the volumetric discharge rate Q, how would you then calculate the mass discharge rate?

You would multiply by the density, since

$$\frac{\text{mass}}{\text{time}} = \frac{\text{mass}}{\text{volume}} \times \frac{\text{volume}}{\text{time}}.$$

22. Consider a vertical cylindrical storage tank that is 40 ft high and 20 ft in diameter, vented to the atmosphere, and filled to the 38-foot mark with diesel fuel (specific gravity about 0.85). Assume that a three-inch diameter hole develops at the bottom of the tank wall where a pipe breaks off right at the tank wall.

a) What will be the initial discharge rate Q?

Using a coefficient of discharge of 0.62, we obtain

$$Q = C_d \times \sqrt{2gh} \times A$$

$$= 0.62 \times \sqrt{2 \cdot 32 \cdot 38} \times \pi \cdot \left(\frac{1}{8}\right)^2$$

$$= 1.5 \text{ ft}^3/\text{sec} = 90 \text{ ft}^3/\text{min} = 674 \text{ gal/min}.$$

b) Letting F represent the total volume of diesel fuel in the tank at any moment, what is the relation between Q and the derivative $\dfrac{dF}{dt}$? Be careful.

The Q calculated above is the negative of the derivative $\dfrac{dF}{dt}$, and the calculation above corresponded to the initial time $t = 0$.

c) Develop the relationships between F, h, Q, and t so that you can finally express F as a function of t. (Here, h is the height of the liquid level in the tank at any time t.)

From the two equations

$$-\frac{dF}{dt} = C_d \times \sqrt{2gh} \times A$$

$$F = \pi \times 10^2 \times h$$

we obtain a simple differential equation for F:

$$-\frac{dF}{dt} = C_d \times \sqrt{2g\frac{F}{100\pi}} \times A$$

$$\frac{dF}{dt} = -.62 \times \sqrt{2 \cdot 32 \cdot \frac{F}{100\pi}} \times \pi \cdot \left(\frac{1}{8}\right)^2$$

$$= -0.014F^{\frac{1}{2}}$$

which, together with the initial condition $F(0) = \pi(10)^2(38) = 11938$, leads to:

$$F^{-\frac{1}{2}}dF = -0.014\,dt$$

$$2\sqrt{F} = -0.014t + C$$

$$2\sqrt{11938} = -0.014(0) + C$$

$$218.5 = C$$

$$F = \left[\tfrac{1}{2}(-0.014t + 218.5)\right]^2$$

$$F = [-.007t + 109.3]^2 = 0.000049t^2 - 1.53t + 11938.$$

d)   If the leak continues unabated, how long would it take for the tank to reach the point where the tank is essentially empty?

Using the equation for F, we have

$$0 = [-.007t + 109.3]^2$$

$$.007t = 109.3$$

$$t \cong 15600 \text{ sec} = 260 \text{ min} = 4 \text{ hrs} 20 \text{ min}.$$

[Note: the result may vary up to about 15900 depending on the degree of precision carried through the previous calculations.]

e)   Assuming the liquid is discharging into some kind of pool, find an expression for the total pool volume as a function of time.

Letting G denote this amount, it is just the "complement" of F:

$$G = 11938 - F = 11938 - [-.007t + 109.3]^2 = -0.000049t^2 + 1.53t.$$

f)   Find an expression for the total mass in the pool as a function of time.

Now we use the density (in pounds per cubic foot) for the only time in this problem:

$$M = .85 \times 62.4 \times G = -0.0023t^2 + 81.2t \quad \text{(pounds)}.$$

23.   Apply your computerized modeling package to the situation in the previous exercise and compare the result with your calculations in part d, above.

Using ARCHIE, we obtain the value 264.5 minutes, which appears to be essentially the same as the answer above, except for differences caused by our rounding of intermediate results.

24.   Suppose that the tank in Exercise 22 is reoriented so that it now is horizontal. Assume that it has the same initial amount of diesel fuel in it as in the earlier situation and that the same size opening occurs at the bottom. How long will it now take to empty?

In this case the geometry is more complex because as the height drops, the cross-sectional area changes. See Figure 7-B. The shaded area (corresponding to an end view of the portion of the tank that contains liquid) is the

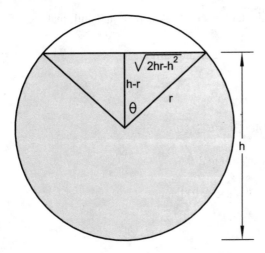 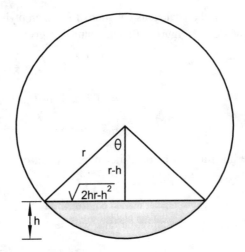

FIGURE 7-B
Geometric formulation of horizontal tank problem (Exercise 24)

sum of the area of the double triangle and the large bottom sector of the circle with angle $2\pi - 2\theta$. The volume is simply this area times the length of the tank. Therefore, the volume F of liquid, at least in the range $h \geq r$, is:

$$F = 40\left[\frac{1}{2}(h-r)(2)\sqrt{2hr-h^2} + \frac{2\pi - 2\theta}{2\pi}(\pi r^2)\right]$$

$$= 40(h-r)\sqrt{2hr-h^2} + 40(\pi - \theta)r^2$$

$$= 40(h-10)\sqrt{20h-h^2} + 4000\left(\pi - \cos^{-1}\left(\frac{h-10}{10}\right)\right).$$

Aside from our further use of this equation below, we can now use a numerical equation solver routine in a calculus package to find the starting value of h, namely the one that gives us the previously determined initial volume of 11,938 ft^3. Thus we solve the equation

$$11938 = 40(h-10)\sqrt{20h-h^2} + 4000\left(\pi - \cos^{-1}\left(\frac{h-10}{10}\right)\right)$$

and find that $h = 18.054$.

The discharge rate is given by

$$-\frac{dF}{dt} = Q = C_d \times \sqrt{2gh} \times A = 0.62 \times \sqrt{64h} \times \pi \times (\tfrac{1}{8})^2 = .2435\sqrt{h}.$$

Therefore, looking at h as the independent variable, the time required to reduce the height of the liquid by an amount Δh would be given by:

$$\Delta t = \frac{\text{incremental volume in this height interval}}{\text{reduction rate applicable to this height interval}}$$

$$= \frac{F'(h)\Delta h}{.2435\sqrt{h}}.$$

Since we earlier made the assumption that $h \geq r$, we can only use this analysis to calculate the time T_1 required to bring the level down to the center-line of the tank. We obtain, using numerical integration,

$$T_1 = \int_{10}^{18.054} \frac{F'(h)}{.2435\sqrt{h}}\,dh = 6332 \text{ sec} = 105.5 \text{ min.}$$

Now we set up the analogous equations for the emptying of the tank once it is less than half full, making reference to the right side of Figure 7-B. The volume is given by

$$F = 40 \left[\frac{2\theta}{2\pi}(\pi r^2) - \frac{1}{2}(r - h)(2)\sqrt{2hr - h^2} \right]$$

$$= 40\theta r^2 - 40(r - h)\sqrt{2hr - h^2}$$

$$= 4000\cos^{-1}\left(\frac{10 - h}{10}\right) - 40(10 - h)\sqrt{20h - h^2}.$$

This time integral will be over the entire interval of h values from 0 to 10:

$$T_2 = \int_0^{10} \frac{F'(h)}{.2435\sqrt{h}}\, dh = 12664 \text{ sec} = 211 \text{ min}.$$

Thus the total time to empty the tank will be:

$$T = T_1 + T_2 = 316.5 \text{ min}.$$

25. Test your result on the previous problem against the corresponding result obtained with your modeling package.

The result from ARCHIE is 315.4 minutes, which shows excellent agreement.

7.2 Pool Size Submodels

1. Discuss any possible limitations you can identify in the logic underlying the first modeling approach given in this section for calculating pool size.

The time dependence of the pool size is completely lost for two separate reasons. First, the use of an average discharge rate in most cases, as was suggested in the text, leaves out the time dependence of the source feeding the pool. Second, use of an equilibrium value ignores the earlier time history of the pool when it will be smaller than this equilibrium value. These effects counterbalance each other to a certain extent because a more rapid initial discharge rate (as, for example, when a tank is more full) will generally be contributing to the pool in its early stages before it reaches its maximum size. One can imagine, of course, a hypothetical time-dependent discharge rate that would lead to a large pool very fast, one that would exceed the size calculated by the given modeling approach. However, for real cases, the approach mentioned errs on the conservative side. Certainly, if the discharge rate is constant, the result is expected to be conservative.

2. It was shown in the text that the units of V are length per time. Give a direct physical interpretation of the V value based on these units, explaining your reasoning.

The units can be interpreted as the rate of decrease in the height of the surface if there were no replenishment by additional liquid moving into the same area. It is easiest to see this in terms of a numerical example. Suppose we had calculated a value

$$V = 0.1 \text{ ft}^3/\text{min per square foot of surface area}$$

which converts to

$$V = 0.1 \text{ ft/min}.$$

The removal of a tenth of a cubic foot from a pool section whose area is one square foot does indeed lower the level by a tenth of a foot, which is exactly what the new units say.

3. The text referred to the assumption that the total vaporization rate should be proportional to the area of the pool. Explain why this is a reasonable assumption, but then also describe physical factors that might cause it to be only an approximation.

Assuming that conditions are uniform throughout the pool, the molecules in one section should not be affected by what happens in another section. Since evaporation is a surface phenomenon, related to the number of molecules in the near surface zone (from which they might escape to the vapor space), if you double or triple the surface area, say, you should have a proportionate doubling or tripling of the number of molecules subject to such escape. The actual fraction escaping per unit of time depends on their energy distribution, which in turn depends on the temperature of the liquid. This has been discussed in Chapter 4.

The net vaporization rate also depends on the rate at which molecules in the vapor space are reentering the liquid phase, and this depends on their concentration in the vapor space, also spoken of previously in terms of their vapor pressure or partial pressure. If a wind is blowing over the surface of the pool, then this will cause the factors to be non-uniform. For example, at the upwind portion of the pool, the vapor space will essentially be "swept out" instantaneously, so the net vaporization rate will be accelerated. On the downwind side of the pool, all the evaporated material from the upwind portion will be in the vapor space already, which may slow down the net vaporization rate there. Even in the absence of a wind, diffusion effects would cause a higher vapor space concentration to accumulate nearer the center of the pool than at the edges.

In addition, there could be temperature variations caused by the wind or by the cooling effect of evaporation (removal of the heat of vaporization), which may have been underway longer in some locations than in others as a result of the pool's time-dependent growth or shrinkage.

These kinds of factors should balance out to a certain extent over the pool, and so the proportionality assumption is still quite good, even if it does not precisely capture these potentially nonlinear effects.

4. Based on the discussion in this section, identify a feedback loop that is inherent in Figure 7-1 but that is not explicitly identified there.

The vaporization submodel has a feedback effect on the pool size submodel, at least in the case when the pool size calculations are based on equilibration of discharge and vaporization.

7.3 Evaporation/Vaporization Submodels

7.3.1 Flash Boiling Submodels

1. Use the above approach to calculate the flash fraction for a spill of liquid propane onto the ground at an ambient temperature of 25°C. Data for propane are: boiling point at atmospheric pressure = –40°C; average specific heat over the temperature range from ambient down to the boiling point = 0.59 cal/g; and heat of vaporization at the boiling point = 102.5 cal/g.

Using the formula in the text, the flash fraction is

$$\text{flash fraction} = \frac{c(u - u_b)}{H} = \frac{(.59)(25 + 40)}{102.5} = .37 = 37\%.$$

2. Read the documentation for your modeling package and summarize the approach taken to deal with the flashing phenomenon.

The ARCHIE model effectively assumes a 100% flash fraction (which includes both gas and aerosols that will soon also change to gas) for a non-refrigerated, low-boiling-point liquid released at ambient temperature.

3. When aerosols evaporate or boil, where does the required heat of vaporization come from?

Since they will have already been brought down to the boiling point during the flashing process, they will have to obtain this additional heat from their surroundings, primarily in the form of conduction from the surrounding air.

4. Discuss the temperature conditions you would expect to find in the air just over a pool of liquid boiling rapidly at ambient temperature. Would this have any implications for the spread of the vapors?

The air will be very cold as heat is removed by the boiling aerosol fraction, as well as from simple heat loss to the cold vapors and the cold pool surface. A fog usually forms over such pools because of the condensation of the water vapor in the air as the air's temperature drops. The cold air mass, being more dense than the surrounding air, would be expected to stay close to the ground. This is discussed further in section 7.4 under the subject of "heavy gases."

5. Describe a physical storage situation involving a low-boiling-point liquid in which a catastrophic tank failure and spill to the ground would not be expected to yield a significant flash fraction.

If the material were stored in a refrigerated tank at a temperature close to its boiling point, then there would be little or no heat content available to support flash boiling. LNG (liquefied natural gas, liquefied methane) is often stored in this fashion. It is handled in such large quantities and its boiling point is so low that at ambient temperature the resulting pressures present significant engineering and cost issues.

7.3.2 Normal Boiling Submodels

1. Verify that $U(x, t)$ satisfies the heat flow PDE for all points with $t > 0$. Furthermore, determine the correct dimensions for κ, u, and U, as used in the calculations above.

Writing the heat flow equation in terms of $U(x, t)$ and then substituting for the derivatives in terms of the specific $u(x, t)$ used to define it, we need to verify that

$$\int_0^x u_t(x, t)\, dx = \kappa u_x(x, t).$$

The left-hand side is an antiderivative with respect to x of the integrand. But so is the right side, for

$$\frac{\partial}{\partial x}(\kappa u_x) = \kappa u_{xx}$$

which equals u_t by the fact that u satisfies the heat flow equation. (This, as discussed in the text, is by analogy with the diffusion equation in Chapter 6.) Since both sides are antiderivatives of the same thing, they can differ only by a constant. However, this is a constant with respect to x, but we still must allow the possibility that it may depend on t. Therefore, we have

$$\int_0^x u_t(x, t)\, dx = \kappa u_x(x, t) + C(t).$$

Now, for an arbitrary but fixed t, look at the point corresponding to $x = 0$. The left side is clearly 0 since it is an integral from 0 to 0. The first term on the right is also 0 because of the factor $(2x)$ in the x-derivative of u. Therefore we have $C(t) = 0$ for all t, and this completes the required verification.

To address the issue of dimensions, by comparing the dimensions of both sides of the heat flow PDE, it follows that κ must have dimensions L^2/t, where L is length and t is time. The equation for u would then imply that u, as written, has dimensions $1/L$. This would make U dimensionless, since its definition involves multiplying u by a length variable, x.

We should note that the dimensions of some of the equations in this development may need to be "adjusted" in order to be interpreted in the desired physical terms. For example, the original u is not in temperature units; but if we multiply by a constant factor with dimensions LT, where T is temperature, then the resulting "adjusted

u" would have the right dimensions. Such a factor would correspond to the M in the concentration equation. This will be seen again in Exercise 3.

The interested teacher may wish to develop this dimensional theme further and verify the reasonableness of the dimensions encountered. However, in this initial introduction to these issues, students often do not fully appreciate the distinctions being investigated.

2. **Determine the boundary conditions satisfied by $U(x,t)$ at $x = 0$ for all $t > 0$ and at $t = 0$ for all $x > 0$.**

This and the following exercise anticipate material that will be worked out in the text later in this chapter. From the definition of this function

$$U(0,t) = \int_0^0 u(x,t)\,dx = 0, \quad \text{for all } t > 0$$

$$U(x,0) = \lim_{t \to 0} \int_0^x u(x,t)\,dx = \frac{1}{2},$$

the last value coming from analogy to the mass diffusion case in Chapter 6, where the limit of the integral should be half the initial mass (this case corresponds to $M = 1$).

3. **Based on your results in the previous two exercises, find a function $U^*(x,t)$ that completely solves the heat flow problem summarized in Table 7-3.**

We form a linear transformation of U of the form

$$U^*(x,t) = aU(x,t) + b$$

and seek appropriate values of a and b to yield the desired boundary conditions. This gives two simultaneous equations in these constants, with the final result being

$$U^*(x,t) = 2(u_g - u_b)U(x,t) + u_b.$$

To verify that U^* satisfies the PDE, you can see by inspection that the b term drops out in both derivatives, and the multiplier a shows up on both sides of the equation and therefore cancels out (unless it is 0, in which case we have a solution anyway, namely, the constant function b).

With respect to dimensions, it was shown in Exercise 1 that U is dimensionless. Therefore, if we think of the constants a and b as having dimensions of temperature, then U^* requires no further adjustment to represent temperature in the specified physical situation.

4. **Discuss the validity of the step in the equations in the text where the partial derivative with respect to t was moved inside the integral sign. (You may wish to consult a calculus book as a reference for this.)**

The usual calculus theorem for differentiation under the integral sign can be stated in the following form: Suppose we have a function $f(y)$ defined on the interval $c < y < d$ by

$$f(y) = \int_a^b g(x,y)\,dx.$$

Suppose further that on the set $\{(x,y)|a \le x \le b, c < y < d\}$, both $g(x,y)$ and $g_y(x,y)$ exist and are continuous. Then the function $f(y)$ is differentiable and its derivative is given by

$$f'(y) = \int_a^b g_y(x,y)\,dx.$$

In the present problem, this amounts to verifying the continuity of our specific function $u(x,t)$ and its time derivative $u_t(x,t)$, which is apparent from the way these functions are built up from continuous functions. (Note that we are not involved with points where $t = 0$.)

5. If $C(x, t)$ is the solution to the standard mass diffusion problem (with mass M initially injected at the origin), give a physical interpretation of the integral

$$m(x,t) = \int_0^x C(s,t)\,ds$$

Using this interpretation together with the diffusion principle, show that m must satisfy the one-dimensional diffusion PDE. (Hint: this is quite simple.)

The integral represents the total mass between the points 0 and x. The time rate of change of this is just the negative of the flux through the boundary at x since by symmetry there will be no flux through 0. Therefore,

$$m_t(x,t) = -q(x,t) = +DC_x(x,t) = Dm_{xx}(x,t)$$

6. **Give two distinct physical interpretations of the function $U(x,t)$ encountered in this section, or of some constant multiple of $U(x,t)$.**

As discussed in the text, the most basic physical interpretation of this function is as the solution to a temperature distribution problem. In particular, following the notation in the solution to Exercise 3, U^* and U will be numerically the same if $a = 1$ and $b = 0$, which corresponds to a situation where $u_b = 0$ and $u_g = \frac{1}{2}$. This implies that U, interpreted in temperature units, gives the temperature distribution underground assuming an initial temperature $U(x,0) = \frac{1}{2}$ for $x > 0$ and a boundary condition $U(0,t) = 0$ for $t > 0$. (We are phrasing this in terms U instead of U^* simply to avoid notational complexity. The concept of interpreting a numerical variable in another set of units is a common abbreviated way of treating the more precise logical step of introducing multiplicative conversion factors with numerical value unity but that are not dimensionless.)

On the other hand, Exercise 5 also suggests that there may be an energy interpretation. In particular, interpreting u in temperature units, if we multiply this temperature function u by the linear density ρ and the specific heat c (assumed to be constant over the temperature range under consideration), then the quantity $\rho\Delta x \cdot c \cdot u(x,t)$ would represent the amount of heat necessary to raise the temperature of a small length Δx of a rod from temperature 0 to temperature $u(x,t)$. But then the integral

$$\rho c U(x,t) = \int_0^x \rho c u(s,t)\,ds$$

would represent the total amount of heat needed to bring the whole portion of the rod between 0 and x from a uniform initial temperature of 0 to its current temperature distribution at time t. (The temperature scale is arbitrary, so the initial uniform temperature could have any absolute value.) Taking into account our specific function $u(x,t)$, with its direct correspondence with the diffusion equation, we know that the integral

$$\int_0^\infty \rho c u(s,t)\,ds = \frac{\rho c}{2},$$

and that as $t \to 0$, this total amount of heat energy is concentrated more and more in a small neighborhood of the end of the rod ($x = 0$). It can thus be seen that u is the temperature distribution and $\rho c U(x,t)$ is the "excess heat content" above the baseline uniform initial temperature condition of the portion of the rod between 0 and x, resulting from the input at time $t = 0$ of an amount of heat equal to $\rho c/2$ heat units at the end of the rod ($x = 0$). (The factor of one-half results from the fact that this is a one-sided problem, corresponding to the equivalent two-sided problem, where an initial heat amount of ρc units is injected, but is allowed to flow both to the left and to the right symmetrically.) It follows by division by ρc that $U(x,t)$ itself, interpreted as representing heat units, represents the excess heat content of the portion of the rod between 0 and x, resulting from the initial input of an amount of heat equal to 1/2 heat unit at the end of the rod ($x = 0$). (The corresponding temperature distribution would be given by $\frac{u}{\rho c}$.)

This latter point of view is pursued further in the text following the current group of exercises.

7. **For this problem, you are given that u, u_1, and u_2 are arbitrary solutions to the heat flow PDE. They do not necessarily have the form of u above, nor do they necessarily satisfy the boundary conditions for our**

problem. Furthermore, a and b are constants. Determine which of the following related expressions are also necessarily solutions to the heat flow PDE:

a) au

b) $u + b$

c) $au_1 + bu_2$

d) $u + bx$

e) u_x

f) u_t

g) u_{xxt}

h) $\int_0^x u(s, t)\, ds$

i) $\int_0^1 u(x, t)\, dx$

j) $\int_{-x}^x u(s, t)\, ds$

All are solutions except part i.

This is reminiscent of issues raised earlier in the exercises in Section 6.4, as well as previously in this section. The expressions in parts a through h are solutions by the arguments given earlier. In particular, for a through d, you can see practically by inspection that when you plug them in to both sides of the PDE it reduces to a known condition. For parts e through g, the result is true as long as the partial derivatives of the requisite order are all continuous, so that the order of partial differentiation can be changed. Part h is as discussed above. Part i is not in general a solution to the PDE. It is a function of t alone, and so the double x derivative will definitely be 0, but the t derivative will generally not be identically 0. (Note that this integral represents the average temperature, at time t, over the interval $0 \leq x \leq 1$. The expression in part j can be rewritten as the sum of two integrals

$$\int_{-x}^x u(s, t)\, ds = \int_0^x u(s, t)\, ds + \int_{-x}^0 u(s, t)\, ds$$

$$= \int_0^x u(s, t)\, ds - \int_0^{-x} u(s, t)\, ds$$

$$= \int_0^x u(s, t)\, ds + \int_0^x u(-r, t)\, dr,$$

so, in light of part h, the question reduces to whether the function $u(-x, t)$ satisfies the PDE. It does, because

$$\frac{\partial}{\partial(-x)} u(-x, t) = u_x(-x, t) \cdot (-1)$$

$$\frac{\partial^2}{\partial(-x)^2} u(-x, t) = u_{xx}(-x, t) \cdot (-1) \cdot (-1) = u_{xx}(-x, t),$$

and so we have finally

$$\kappa u_{yy}(y, t) = u_t(y, t) \qquad \text{(given)}$$

$$\kappa \frac{\partial^2}{\partial(-x)^2} u(-x, t) = u_t(-x, t) \qquad \text{(change name of variable)}$$

$$\kappa u_{xx}(-x, t) = u_t(-x, t)$$

which verifies the PDE.

8. For our original problem involving a pool of boiling liquid, is this last condition of zero heat flux at $x = 0$ satisfied?

No, since heat is flowing through the boundary into the liquid pool.

9. For the function u defined by

$$u = \frac{1}{\sqrt{4\pi\kappa t}} e^{-\frac{x^2}{4\kappa t}},$$

is the zero heat flux condition met at $x = 0$? (Hint: this should be very easy to see.) Relate your result here to your solution to Exercise 1.

Yes, since the derivative with respect to x is 0 at $x = 0$, and the flux is proportional to this derivative. Alternatively, the symmetry of this system implies 0 flux at the origin.

The relationship is the following. We verified in Exercise 1 that U, defined as an integral of the u in this problem, satisfied the PDE. By the text preceding this set of exercises, key to this verification would have to be use of this zero-flux condition (or else the result could not be true). Looking back on the solution to Exercise 1, the solution did indeed make use of the fact that $u_x(0, t) = 0$, which is equivalent.

10.   Other than for the situations investigated in the previous exercises, describe a physical situation of one-dimensional heat flow in which there would be a zero-flux boundary.

A heat conduction rod with an insulated end, so that there could be no heat flux through it, would be a simple example.

11.   Verify the limit of 1/2 in the previous equation. (Hint: you should be able to do this practically by inspection based on the work in the previous chapter or on any previous experience you may have had in statistics.)

As pointed out in the solution to Exercise 2, this can be interpreted as half the injected mass in a mass diffusion problem with $M = 1$, and hence has the value $1/2$. Alternatively, if you change the variable of integration to $z = \dfrac{s}{\sqrt{2\kappa t}}$, the limit reduces to half the area under the standard normal distribution and hence has the value $1/2$.

12.   As an alternative strategy to the previous problem, you might argue that since the integral is taken with respect to s, terms involving t can be moved in or out of the integral to suit your convenience, just like constants. Therefore you could take the factor in front of the integral inside, and then also the limit. For this line of argument, either justify it based on the rules of calculus and use it to get the correct answer, or determine precisely where it violates a rule of calculus for taking constants and limits inside an integral.

Moving the quotient factor inside the integral sign is indeed just like moving a constant inside, and this is a valid operation. However, moving the limit inside is not valid in this case. For example, having moved the quotient factor inside, we have the limit:

$$U(x, 0) = \lim_{t \to 0} \int_0^x \frac{1}{\sqrt{4\pi\kappa t}} e^{-\frac{s^2}{4\kappa t}} \, ds.$$

The value of x may be regarded as fixed, and so this is involves a normal Riemann integral, not an improper integral. Thinking of the integrand as a concentration function associated with a point of mass injection at the origin, or its heat flow equivalent, this family of continuous functions, as $t \to 0$, does not even converge at $x = 0$, is thus certainly not uniformly convergent, and is not even uniformly bounded because they have higher and higher peak values $\dfrac{1}{\sqrt{4\pi\kappa t}}$ at the origin and hence also in small intervals adjacent to the origin. This violates typical convergence theorems from calculus, the most common of which is that uniform convergence of the integrand to its limit function implies that the limit can be taken inside the integral.

Incidentally, the limit of the integrand is 0 for $x > 0$, as can be seen either from the application of L'Hospital's rule, or from consideration of the concentration interpretation described above.

13.   Set up and solve the two simultaneous equations referred to in the previous paragraph.

(This would have been done in connection with Exercise 3 as well.) The equations are:

$$B = u_b$$
$$\frac{A}{2} + B = u_g.$$

The first already gives the B value, and substitution of this into the second gives $A = 2(u_g - u_b)$.

14. **Show that U^* satisfies the heat flow PDE, based on the fact that U does.**

This follows from Exercise 7, parts a and b, or, alternatively, from the argument given in the solution to Exercise 3.

15. **Draw a series of three graphs, corresponding to progressively later times, on one set of axes to represent qualitatively the expected temperature profile going down into the ground under a liquid pool boiling at ambient temperature. Then, on a set of axes just below and lined up with the first, draw graphs of the temperature gradients for the original three graphs. Explain how you can use these graphs to illustrate some of the concepts from this section concerning the boiling rate as a function of time.**

See Figure 7-C. Early in the development of the incident, heat is drawn from the nearby portions of the underground. The temperature gradient is very high, and heat flows from nearby relatively rapidly upward to the pool. As time progresses, the sharp initial gradient near the surface smoothes out, and the temperature shows a more gradual increase proceeding downward. The temperature profile levels out further as time progresses, and the resulting lower gradient causes less heat to flow into the pool. The net result is a gradual slowing of the boiling rate.

16. **Suppose that a refrigerated tank containing 1 million kilograms of LNG (liquefied natural gas) fails catastrophically and that the liquid forms a pool in the diked area, which contains 20,000 ft^2. How long would it take for all of the mass to enter the vapor cloud? What would be the flash fraction? [You may assume that the initial temperature of the contents of the tank is –162°C, the boiling point of the material. The heat of vaporization of LNG at this temperature is 119 cal/g. For other parameters, use the following typical values: ambient temperature 25°C; initial ground temperature 18°C; ground specific heat 0.2 calories per gram per**

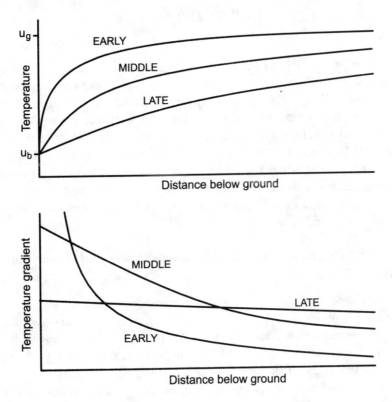

FIGURE 7-C
Temperatures and temperature gradients as a function of distance underground, for various times (Exercise 15)

degree Celsius; ground thermal diffusivity 0.0046 cm^2 per second; and ground density 2.5 grams per cubic centimeter.]

To begin, the flash fraction is obviously 0 since the material is being released at its boiling point, not above it. The various relevant parameters are:

$$u_g = 18°C$$

$$u_b = -162°C$$

$$H = 119 \text{ cal/g}$$

$$\kappa = .0046 \text{cm}^2/\text{sec}$$

$$\rho = 2.5 \text{ g/cm}^3$$

$$c = .2 \text{ cal/g-}° \text{ C}$$

$$K = \kappa\rho c = .0023 \text{ cal/cm-sec-}° \text{ C}$$

$$A = 20000\text{ft}^2 = 1.858 \times 10^7 \text{ cm}^2$$

$$M = 10^6 \text{ kg} = 10^9 \text{ g.}$$

(The ambient temperature is not relevant.) Therefore, using the final equation derived in the above text, we can solve for the time t as follows:

$$\frac{K(u_g - u_b)}{H\sqrt{\pi\kappa}}(2\sqrt{t} - 2\sqrt{0}) \times A = M$$

$$\frac{(.0023)(180)}{119\sqrt{\pi(.0046)}}(2\sqrt{t}) \times 1.858 \times 10^7 = 10^9$$

$$(1075413)\sqrt{t} = 10^9$$

$$t = 864670 \text{ sec} = 10 \text{ days.}$$

This value may seem surprising, but it is correct based on the model we are using. The precise value is not so important, but it serves to indicate that this emergency would exist over a relatively long period of time and that emergency planning measures could be quite valuable in bringing it under control. The biggest weakness in the model is the lack of consideration of heat flow from other sources once the reservoir of heat in the ground near the surface has been substantially depleted. For example, heat could move laterally from the underground area not immediately under the pool. This could be rectified by using higher-dimensional heat-flow modeling. The result also suggests that heat transfer by conduction from the air or by solar radiation, might also become important. (Of course, the most rapid vaporization would occur early and would present the highest concentration levels in the plume. Therefore the maximum hazard distance is not much affected by these limitations, and this is probably the most important value one would want to calculate from a model.)

17.   Compare your results in Exercise 16 with the same results obtained using your modeling package. Can you explain any differences?

The ARCHIE package yields a result of several hours, depending on assumptions about weather conditions. This is because heat transfer is not incorporated as a limiting factor. This is an example of the application of two modeling approaches (this one and the one discussed above), each being "stretched" beyond its envelope of reliable applicability. The multiple conservatisms built into ARCHIE provide marginally useful conservative results in this kind of case, and it is good to recognize that this can happen.

(Although it would be beyond the scope of this exercise, one could focus a comparison on the earliest stages of the incident, when heat is more readily available to the pool, and the vapor evolution rates would be expected to be more similar.)

18. **Suppose you were to repeat Exercise 16 with the only change in input data being an increase in the ground density to 4 grams per cubic centimeter. Qualitatively speaking, how should this affect the results, if at all? Explain.**

This would increase the heat content of the ground, and therefore more heat would be available for transfer from any point into the pool. This would increase the boiling rate and decrease the total time required for boil-off.

19. **Suppose you were building a modeling package to apply to liquids with low boiling points. Draw a general logic diagram or flowchart to illustrate how you would combine the following five submodels: discharge, pool formation, flash boiling, normal boiling, and vapor dispersion.**

A reasonable structure is shown in Figure 7-D. The nature of the feedback from the boiling or vaporization submodel to the pool-size model is the most problematic component since the boiling rate depends on time. One approach would be to run through the model one time-step at a time, letting the pool increase on a time-step if the new input exceeds the amount vaporized in the last time-step. This would require some relationship between area and volume of the pool. Another approach would be to adopt an average boiling rate to use over a whole set of time intervals, which would take account of the fact that the heat in the ground right near the discharge will be

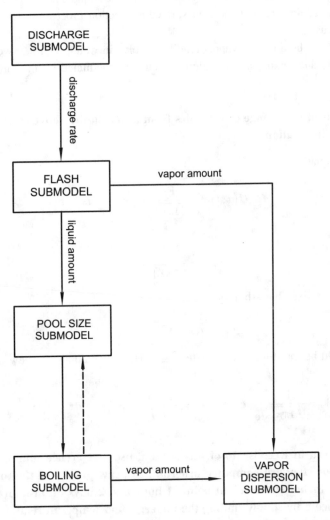

FIGURE 7-D
Submodel relationships for low-boiling-point liquids (Exercise 19)

further depleted at any given time than in more remote locations that are reached by the growing pool only after some period of time has elapsed.

20.   Consider spills of low-boiling-point liquids onto bodies of water instead of onto land, such as if a tank barge carrying such material were rammed by another vessel, breaching its tanks while it is moored in a harbor. What would you expect to be the differences in the evolution of the incident compared to a spill on land? You may assume that the material is lighter than water and largely floats as a pool on the surface.

In general, vaporization would occur faster for a number of reasons.

Water has a higher specific heat than most substances, which causes it to offer a greater heat reserve to support boiling. This is true even though its density is less. For example, the specific heat of water is 1 calorie per gram per degree Celsius, whereas the corresponding value for average soil is about 0.2 . Soil has a density of about 21/2 times that of water, but this still raises its heat content, on a per volume basis, to half of that for water.

Furthermore, heat can move by convection in a body of water, rather than only by the very slow conduction process available in solids. Colder water (at least down to 4°C) will sink, and warmer water from below will rise to replace it due to buoyancy. This accelerates the flow of heat to the liquid pool floating on the surface of the water.

In addition, turbulence caused by boiling will tend to mix the top layer of the water so that its heat can be transferred quickly.

When the water reaches its freezing point, this also frees up considerable heat (80 cal/g) for the boiling process, although the formation of ice does restrict the near-surface convection processes. But ice still does conduct heat about twice as effectively as soil.

On the other hand, if the liquid or its vapors readily dissolve in water, then this would serve to remove some of the potential vapor risk, and mixing and dissolution would be facilitated by the turbulence of the boiling and flashing processes.

21.   Work out the details of the change of variables from s to z in the above calculations, including consideration of the bounds of integration.

Beginning with the equation

$$U(x,t) = \frac{1}{\sqrt{4\pi\kappa t}} \int_0^x e^{-\frac{s^2}{4\kappa t}} ds$$

and the change of variable

$$z = \frac{s}{\sqrt{4\kappa t}},$$

we obtain, thinking of t as a fixed but arbitrary value,

$$ds = \sqrt{4\kappa t}\, dz$$

and the upper bound would be transformed from x to $\frac{x}{\sqrt{4\kappa t}}$. Hence we obtain the new integral

$$U(x,t) = \frac{1}{\sqrt{4\pi\kappa t}} \int_0^x e^{-\frac{s^2}{4\kappa t}} ds = \frac{1}{\sqrt{\pi}} \int_0^{x/\sqrt{4\kappa t}} e^{-z^2} dz = \frac{1}{2}\mathrm{erf}\left(\frac{x}{\sqrt{4\kappa t}}\right).$$

22.   For the problem given in Exercise 16 of this section, use a reference source of error function values (erf z), such as a table or standard computer program, to draw graphs of the underground temperature profile within the top 20 inches of the soil at points 1 hour and 2 days into the process. (Hint: If you have trouble getting started, begin by simply finding the underground temperature at a single time and location, say 1 hour into the process at a location 2 inches underground. This should help clarify how to do the required calculations.)

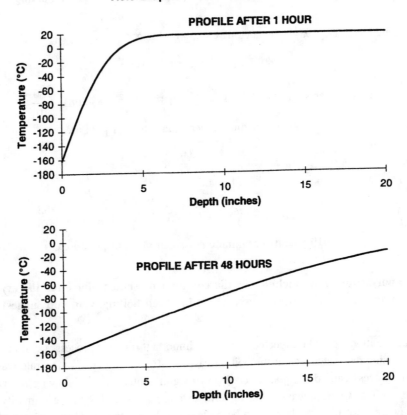

FIGURE 7-E
Temperature profiles for Exercise 22

The underground temperature profile was shown earlier to be given by

$$U^*(x,t) = (2u_g - 2u_b)U(x,t) + u_b$$

which in this situation becomes

$$U^*(x,t) = 360U(x,t) - 162$$
$$= 360\left[\frac{1}{2}\mathrm{erf}\left(\frac{x}{\sqrt{4\kappa t}}\right)\right] - 162$$
$$= 180\,\mathrm{erf}\left(\frac{x}{\sqrt{4\kappa t}}\right) - 162.$$

For $\kappa = .0046$ and for each of the two times, 1 hour and 2 days, we use standard reference values of this function to construct the graphs shown in Figure 7-E. (Of course, we first convert these times to the units of the problem, namely, seconds, and we apply this equation with distances in centimeters.)

7.3.3 Evaporation Submodels

1. **What would be the estimated evaporation rate of acetone from a spilled pool at 20°C? Assume a wind speed of 5 mph. Find the corresponding rate for acrylonitrile, a chemical also treated in earlier examples.**

Applying the equation in the text to the acetone case, we obtain:

$$E = 4.65 \times 10^{-6} u^{0.75} (1 + 4.3 \times 10^{-3} T^2) \cdot M_w \cdot \frac{P}{P_h}$$

$$= 4.65 \times 10^{-6} (5)^{0.75} (1 + 4.3 \times 10^{-3} (20)^2) \cdot (58.08) \cdot \frac{180}{10}$$

$$= .044 \text{ pounds per minute per square foot of pool area.}$$

Similarly, for acrylonitrile, we have:

$$E = 4.65 \times 10^{-6} u^{0.75} (1 + 4.3 \times 10^{-3} T^2) \cdot M_w \cdot \frac{P}{P_h}$$

$$= 4.65 \times 10^{-6} (5)^{0.75} (1 + 4.3 \times 10^{-3} (20)^2) \cdot (53.06) \cdot \frac{83}{10}$$

$$= .019 \text{ pounds per minute per square foot of pool area.}$$

2. The heat of vaporization is not included in the equation presented above for the evaporation rate, and yet it was quite important in the previous case, which involved boiling. Can you suggest any possible basis for this?

The heat of vaporization would be expected to be most important if (i) the process was limited by the availability of heat flow from outside the system (as in the boiling case), or (ii) if the vaporization rate was so substantial that the heat lost by this process caused the pool to cool down significantly, thereby slowing the vaporization process. Although this second effect would always occur to some extent, such heat losses are primarily near the surface of the pool, where they are most likely to be able to be replenished by contact with the ambient air.

Nevertheless, one might still expect the heat of vaporization to show up somewhere in the evaporation equation since, regardless of the source of the heat (e.g., the latent heat in the liquid or heat flowing in from outside the system), a certain amount is necessary to vaporize every single unit of mass that evaporates. Therefore, other factors being equal, one would expect a material with a higher heat of vaporization to have a lower evaporation rate. This aspect is actually represented indirectly in the evaporation equation, because, as should appear quite reasonable, there is a strong inverse correlation between vapor pressure (which is in the equation) and heat of vaporization. (For further details on this relationship, one could consult a physical chemistry text.)

7.4 Vapor Dispersion Submodels

1. A certain pool of acetone is estimated to be evaporating at a rate of 50 lb/min into the air. The conditions are bright sun with a wind at 8 mph. What would the ground-level concentration be at a distance of 350 feet downwind along the direct axis of the wind? Express your final answer in units of grams per cubic meter in order to facilitate comparisons with the next exercise. (Do this problem using your air dispersion tools from Chapter 3, not your hazmat modeling package.)

This problem provides an example of the frequent unit changes one encounters in applying modeling packages; it is simply a fact of life that one must live with. Using an air dispersion spreadsheet program (as in Chapter 3) and atmospheric stability class B, we obtain a concentration of 0.14 g/m^3.

2. Repeat the above problem using your modeling package. How do the results compare?

We use the toxic vapor dispersion component of ARCHIE, even though we have no specific toxic threshold that we are measuring against. The output of this module includes tables with concentration as a function of downwind distance. Furthermore, the outputs are in ppm units, but if you put in a concentration threshold of, say, 1 g/m^3, the program converts it to ppm units and basically gives you the right conversion factor to use to reconvert the answers to units of g/m^3 for comparison with the results of the previous problem.

Allowing a long release duration so that the model simulates a plume, we find from the output tables by interpolation that the concentration at 350 feet downwind would be about 66 ppm, which converts to 0.16 g/m^3. This is about 15% higher than the value calculated in the previous exercise. Review of the ARCHIE documentation would indicate that the program uses a variant of the Gaussian plume model, thus accounting for some variation from the results obtained in the earlier exercise. (But agreement to within 15% is quite good and well within the overall range of uncertainty in such models and data values. Just consider the effect of a one-mph variation in windspeed, for example, whose corresponding dilution factor would be of this same magnitude.)

3.   Write a simplified version of the Gaussian plume equation for the case of a ground-level release and a ground-level receptor. (This will be our most common application.)

In this case both H and z are 0, so from the original Gaussian plume equation

$$C = \frac{Q}{2\pi\sigma_y\sigma_z u}\left[e^{-\frac{y^2}{2\sigma_y^2}}\right]\left[e^{-\frac{(z-H)^2}{2\sigma_z^2}} + e^{-\frac{(z+H)^2}{2\sigma_z^2}}\right]$$

we obtain

$$C = \frac{Q}{\pi\sigma_y\sigma_z u}e^{-\frac{y^2}{2\sigma_y^2}}.$$

4.   Discuss the last bracketed factor, involving the z-direction, in the Gaussian plume equation, explaining why it accounts for a no-flow boundary condition at the surface of the ground. Does this apply even when $H = 0$, or does it double the correct value in this situation?

As discussed in Section 6.3, this term includes an imaginary or "virtual" source at a distance H below the ground. The symmetry between the real and the virtual source causes a no-flow boundary to occur at the ground level so that the releases from the real source are always modeled as being confined to the space above the ground, which is what one would want. When $H = 0$, the value given by the equation is still correct. In the real case, all the material would stay within the area above the ground. The model makes it appear as though there is twice as much source material, but it counterbalances this by mathematically allowing half of it to diffuse into the imaginary space below the ground.

5.   Discuss the rationale for excluding dispersion in the x-direction in the Gaussian plume model. Is this a conservative assumption in terms of calculated values of maximum downwind concentrations?

The rationale for this assumption is that since diffusion (or dispersion) is driven by a concentration gradient, it should be much greater in the y- and z-directions. For, in the steady-state situation covered by the Gaussian plume model, there is only a very gradual change (i.e., very small gradient) in concentrations in the x-direction, whereas there is a much more rapid drop-off in concentrations extending from the plume centerline outward in the y- and z-directions.

This would generally be expected to be a conservative approach, at least with respect to most issues related to maximum concentration levels. The rationale would be that diffusion always moves material from areas of higher concentration to lower concentration, so it tends to smooth things out and hence diminish peaks.

6.   If the combination of air and vapor is heavier than the air itself, would it be conservative to ignore this difference and apply the Gaussian plume model?

No, since ground-level concentrations would be higher if this effect decreased vertical diffusion upward or, in the case of an elevated source, effectively moved the plume axis downward towards the ground.

7.   How does the assumption cited above relating to a long time of release relate to any of the other assumptions on the list? Describe qualitatively the effect on concentration values of dropping this assumption in favor of a constant, but finite-duration release.

The long time of release assumption implies a steady source, which decreases concentration variations with time, and hence also with distance downwind. This small axial concentration gradient is the basis for the subsequent assumption that diffusion in this direction can be ignored. For a constant but finite release, axial dispersion near the beginning and ends of the plume could be considerable because the concentration gradients will be large there. This will smooth out concentrations, lessening magnitudes, but causing the plume to arrive earlier and depart later at any distance downwind. (Note: since the idealized diffusion model assumes that mass is spread out an infinite distance, even if the concentration is very low, "arrive" and "depart" would need to be defined in terms of a threshold concentration level.)

8.   Suppose your σ_y and σ_z values do indeed pertain to ten-minute averages, but you interpret the model results as instantaneous concentration values. Have you overestimated or underestimated the concentrations you wanted? What if you interpreted your results as half-hour averages of concentration values?

You have underestimated the dispersion rate for the instantaneous case and overestimated it for the half-hour case. Longer time averages, just like larger samples in statistics, tend to have a lower variability. Therefore, the σ_y and σ_z values for longer-time averages would be expected to be lower. However, the effect of this on concentrations depends on where you are calculating such concentrations. Less dispersion means lower concentrations away from the plume centerline, especially near the source, and hence greater concentrations at longer distances along the centerline and near to it, since the plume holds together for a longer distance. (You can also see this mathematically by looking at the way the σ's enter the dispersion equation in both the front coefficient and in the exponent.)

9.   How might you adapt the Gaussian plume model to a situation where the release rate varies with time? Does your approach apply to the situation where the release rate includes a fairly steep rise? If so, explain how. If not, can you find any additional way to include this situation?

There are various ways to do this. If the source is slowly varying, then, as discussed earlier, dispersion in the x-direction might still be overlooked. Then the concentration at any point at a given time could be calculated as the concentration that would result from the Gaussian plume model applied with the source strength applicable at the moment that that section of the plume passed by the source. (That is, there would be time delay equal to the downwind distance divided by the wind speed.) Alternatively, if the model has been adapted to a finite duration but constant source, it could be further adapted to this situation by approximating the current source function by a sum of constant, finite duration sources of different durations.

But the real area of difficulty is in modeling rapid changes in source strength, such as at the beginning of an event. In this case, depending on the modeling objectives, one might need to incorporate axial dispersion. One way to do this is to look at the plume as a sequence of short puffs (as discussed earlier) and to superimpose the resulting concentrations calculated from these. Another approach would be to look at the plume front as a one-sided puff, meaning one that could diffuse downwind but not upwind (because the gradient upwind would be practically zero due to the steadiness of the plume, once it got started). This would be just like having a no-flow boundary on the y, z-plane through the center point and could be treated using the reflection principle. The puff model is further pursued in the text and the next set of exercises.

10.   Give a convincing heuristic explanation for the puff equation presented above. Devote particular attention to a simple explanation for the first bracketed term, involving the x-direction, and to the determination of the composite term in front of the bracketed ones. You may use any results from earlier chapters in your explanation.

In Exercise 23 of Section 6.1, we developed the basic puff equation for a release into three-dimensional space. This equation was:

$$C = \frac{M}{(4\pi t)^{3/2}\sqrt{D_1 D_2 D_3}} e^{\left(-\frac{x^2}{4D_1 t} - \frac{y^2}{4D_2 t} - \frac{z^2}{4D_3 t}\right)}.$$

The constant term in front followed from the fact that the integral of the concentration over 3-space had to be the total source amount. In Exercise 6 of Section 6.4, we further modified this to model a release above the ground, using a no-flow boundary at the ground. The equation was given as:

$$C = \frac{M}{(4\pi t)^{3/2}\sqrt{D_1 D_2 D_3}} e^{\left(-\frac{x^2}{4D_1 t} - \frac{y^2}{4D_2 t} - \frac{(z-H)^2}{4D_3 t}\right)} + \frac{M}{(4\pi t)^{3/2}\sqrt{D_1 D_2 D_3}} e^{\left(-\frac{x^2}{4D_1 t} - \frac{y^2}{4D_2 t} - \frac{(z+H)^2}{4D_3 t}\right)}$$

which can easily be rewritten in the form:

$$C = \frac{M}{(2\pi)^{3/2}\sqrt{2D_1 t \cdot 2D_2 t \cdot 2D_3 t}} \left[e^{-\frac{x^2}{4D_1 t}}\right]\left[e^{-\frac{y^2}{4D_2 t}}\right]\left[e^{-\frac{(z-H)^2}{4D_3 t}} + e^{-\frac{(z+H)^2}{4D_3 t}}\right]$$

$$= \frac{M}{(2\pi)^{3/2}\sigma_x \sigma_y \sigma_z} \left[e^{-\frac{x^2}{2\sigma_x^2}}\right]\left[e^{-\frac{y^2}{2\sigma_y^2}}\right]\left[e^{-\frac{(z-H)^2}{2\sigma_z^2}} + e^{-\frac{(z+H)^2}{2\sigma_z^2}}\right]$$

where the σ's replace the obvious factors involving the diffusion coefficients and time. But now, when we add a wind factor, the entire puff basically migrates in the x-direction at a speed equal to the windspeed u, so by time t its center is at $x = \mu t$, and therefore the x-distance from the center is $x - \mu t$. This replaces x in the above equation and yields the desired final form. The σ's now take on their usual meaning as functions of distance, which is a surrogate for time in the previous formulation.

11. **Use the puff equation to model the concentration of dangerous ethylene oxide vapor resulting from an instantaneous nighttime release of 50 lb of such vapor from a vent that is 30 feet above ground level. For a ground-level receptor at a distance of 400 feet downwind, directly along the axis of the wind, find both the maximum concentration and the concentration at a time 30 seconds after the release. Assume that the wind is blowing at 10 mph and that it is a clear night. Also assume open terrain, as usual, and use the dispersion coefficients from Chapter 3 (even though they are really intended for time-averaged concentrations). If you need any additional input parameters, make reasonable assumptions for their value(s), explaining your rationale.**

We convert everything to metric units and determine the y and z dispersion coefficients either from the graphs in Chapter 3, the numerical approximations given in that chapter, or from our air-dispersion spreadsheet program. The only key input that is lacking is the dispersion coefficient in the x-direction. Since this is also horizontal dispersion, we use the same value as for dispersion in the y-direction. (This is discussed further in the text after these exercises.) The maximum concentration at our receptor location will occur, of course, when the center of the puff passes over the point, which is theoretically at 27.3 seconds (distance divided by wind speed). This value leads to a theoretical concentration of 2.0 g/m³. Just 2.7 seconds later, the concentration has dropped to 0.5 g/m³. The input parameters, intermediate results, and the final result for this second calculation are shown below in a table from a simple spreadsheet constructed for this problem:

M	22,727.27	coef	5.270405
sigx	7.4	xterm	0.252925
sigy	7.4	yterm	1
sigz	5	1st zterm	0.187823
u	4.473	2nd zterm	0.187823
H	9.144	sum zterms	0.375646
x	121.92		
y	0	C=	0.500743 g/m3
z	0		
t	30		

12. **Apply your modeling package as best you can to the previous problem and compare the results with your previous calculations. Discuss any observations.**

ARCHIE does not calculate time-dependent concentrations except in terms of arrival and departure times of high concentrations at various distances. However, simulating the instantaneous release as a 6-second release (the shortest permitted by the model) at a rate of 500 pounds per minute (for a total of 50 pounds), it calculates a peak concentration at our receptor location of about $0.8 g/m^3$ (converted from the ppm value interpolated from the output table), and an overall peak ground-level concentration of $1.1 \ g/m^3$ at a point 564 feet downwind. Since we observe a high sensitivity to short changes in time, it is not surprising that the extended release required for ARCHIE does diminish the peak value predicted for the receptor location.

One should not look at these results and say automatically that ARCHIE has more limitations when it comes to the puff situation. We are not really interested in approximating the idealized puff situation, but rather we are using various approaches, such as puff and plume modeling, to try to model real situations, which always unfold over a finite length of time. The sensitivity to time mentioned above can be reinterpreted to say that for short, finite duration releases, the puff model can cause considerable error because it does not account for this finite extent and concentrates all the mass at a single point in space and time. (A clever way to improve on this latter situation is to represent a finite extent source by a "virtual puff" or "virtual source" at an earlier point in time and upwind from the real source, so that by the time the idealized puff reaches the source location and real starting time, its geometry is more similar to the real source.)

13.   Suppose you tried to model the situation in Exercise 11 by using the standard Gaussian plume model, assuming a release rate of 50 pounds per second. (That is, you use a brief time period to approximate an instantaneous release, so that now you have a discharge rate.) Find the predicted concentration at the same point as above, and discuss your results.

The receptor concentration has no time dependence with this model since the source is assumed to be steady. The result from applying our standard air dispersion spreadsheet is a concentration of $8.4 \ g/m^3$ at the receptor. This is four times larger than the value from the puff model, primarily because there is no axial dispersion to lower the value more rapidly.

14.   In previous exercises you have had to make assumptions about σ_x and other parameters. Modelers need to examine the sensitivity of their results to such assumptions, especially when there is considerable uncertainty about them. Returning to the situation of Exercise 11, consider the answer to the second question there (i.e., the concentration at a fixed point at $t = 30$) as a function $f(\sigma_x)$. You should be able to calculate its value for a range of σ_x values nearby (although it will be easier if you program the puff equation into a programmable calculator, spreadsheet, or other computer program). Draw the graph of the function $f(\sigma_x)$ for a reasonable domain of values for σ_x. What does this graph tell you about the sensitivity of the puff calculations to this parameter? (E.g., is it large or small?)

The requested graph is shown in Figure 7-F. This shows that the concentration value varies by at most a factor of 2 as this coefficient ranges between 6 to 8. Given many of the other large uncertainties associated with this modeling problem, this would not appear to be an excessive amount of sensitivity. If the slope of this curve were very steep, so that small changes in σ_x could change the result by an order of magnitude or more, then there would be greater need to pin down the precise σ_x most applicable to this particular problem. This kind of "sensitivity analysis" is intended to clarify the main sources of uncertainty in model results, and then it is up to the person who will be using those results to determine whether such uncertainties are acceptable or whether further modeling is needed.

15.   Consider a vapor release from a pool of liquid acrylonitrile with a windspeed of 5 mph and atmospheric stability class D. Suppose the pool is growing in such a way that the rate at which material enters the vapor state increases linearly from an initial rate of 100 lb/min to a rate of 600 lb/min 30 minutes later. At that point the fire department covers the pool with foam, thereby stopping evaporation. You are interested in knowing the concentration at this point in time (i.e., when the foam cover is finally established) at a location that is 350 feet downwind and 40 feet to the side of the axis of the wind through the pool center. Express

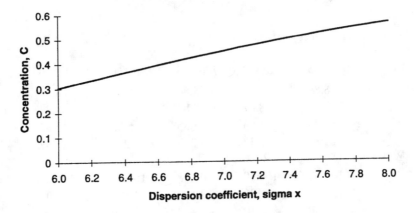

FIGURE 7-F

Sensitivity analysis for concentration at a fixed location and time with respect to σ_x (Exercise 14)

your answer in grams per cubic meter to facilitate later comparisons. Do this problem along the lines just discussed in the text. In particular:

a) Model the release as 6 five-minute puffs.
b) Model the release as 60 half-minute puffs.
c) Based on this experience, construct what you believe to be a reasonably good approximation to the answer.
d) How would your solution method change if you wanted the concentration at the 20- or 40-minute marks? (You do not need to work out the actual numerical values for these cases.)

[Hint: in order to make this problem computationally tractable, you will need to use a spreadsheet program (probably the best for this), other computer program, or programmable calculator.]

We begin by accumulating all the necessary data parameters and converting them to metric units. We will use the same puff spreadsheet that we developed for Exercise 11. In this case the source function has the form

$$m(t) = 100 + \frac{t}{30 \cdot 60}(500) = 100 + \frac{5t}{18}$$

where t is now in seconds (although $m(t)$ is still in pounds per minute). If we divide the interval into 6 components and use the midpoint of the time interval as the representative time and the integral of the source function as the corresponding mass (midpoint value times length of interval), we obtain the values shown below:

t(min)	t(sec)	M(pounds)	M(grams)
2.5	150	708.3	321,973
7.5	450	1,125.0	511,369
12.5	750	1,541.7	700,765
17.5	1,050	1,958.3	890,160
22.5	1,350	2,375.0	1,079,556
27.5	1,650	2,791.7	1,268,952

Now we apply our puff model to each of these, but for different times. For example, the first release would be at a nominal time point of 2.5 min, so the duration before the measurement point should be 27.5 min or 1650 sec. The results are shown in Table 7-A and should be very instructive, much more so than if they turned out to be greater than 0! By looking at the $x\text{-}ut$ values, you can see that the representative time points lead to distinct puffs all of whose centers are quite far away from the receptor location at the moment of measurement. After all, at a wind speed of 5 mph (2.237 m/s), the travel time from the pool to the receptor is about 48 seconds, not much time for diffusion, so a puff initiated any significant time away from 30 minutes less 48 seconds is not likely to make a big contribution to the desired measurement, which is at the 30-minute mark. You can see in the far right column of

TABLE 7-A

Initial results on Exercise 15 with six time intervals

M	321,973	511,369	700,765	890,160	1,079,556	1,268,952
sigx	8.84	8.84	8.84	8.84	8.84	8.84
sigy	8.84	8.84	8.84	8.84	8.84	8.84
sigz	5.85	5.85	5.85	5.85	5.85	5.85
u	2.237	2.237	2.237	2.237	2.237	2.237
H	0	0	0	0	0	0
x	106.68	106.68	106.68	106.68	106.68	106.68
y	12.192	12.192	12.192	12.192	12.192	12.192
z	0	0	0	0	0	0
t	1650	1350	1050	750	450	150
x-ut	-3584.37	-2913.27	-2242.17	-1571.07	-899.97	-228.87
coef	44.72	71.02	97.33	123.63	149.94	176.24
xterm	0	0	0	0	0	2.8E-146
yterm	0.386	0.386	0.386	0.386	0.386	0.386
1st zterm	1	1	1	1	1	1
2nd zterm	1	1	1	1	1	1
sum zterms	2	2	2	2	2	2
C=	0.000	0.000	0.000	0.000	0.000	0.000

TOTAL CONCENTRATION = 0.00000

the table that the last puff, the one nominally released at the 27.5-minute mark, is just beginning to get on scale, even if the value is extremely small.

Now we repeat this same process using 60 puffs, each of half-a-minute duration. This involves a simple modification to the previous spreadsheet, and the result is a total concentration of $11.4 \ g/m^3$. However, although the detailed spreadsheet is not shown here, the only non-negligible contribution to the sum is from the next to the last puff, at 30 minutes less 45 seconds!

As a further improvement on this process, we will now divide the time interval up into one-second increments, but we will reduce the computation by leaving out those portions of the time interval which seem to make no significant contribution to the final result. Therefore, let us concentrate only on the thirty one-second intervals from 29 minutes to 29.5 minutes. If we carry out the same kinds of calculations for this problem, the resulting concentration has the value $4.73 \ g/m^3$. A portion of the output table is shown as Table 7-B, where it can be seen that now we have achieved results that are more reasonably distributed and do not suffer from the shortcomings of the previous two cases. In fact, with respect to the previous case (half-minute time intervals), we would now suspect that the value there was too high because the entire mass distribution for a 30-second period happened to have been concentrated into a puff at a time value, 30 minutes less 45 seconds, that had its center point almost right over the measurement point at the time of interest.

If you wanted the concentration at the 20-minute mark, then the key time interval would be around 19 to 19.5 minutes. Furthermore, releases after 20 minutes would not even be relevant. If you wanted the concentration at the 40-minute mark, you should be able to see that it will essentially be 0 since the last puff center will have passed 10 or more minutes earlier.

16.   The approach of the text and the previous problem clearly suggests a limiting process. That is, think of a finite duration plume as a sequence of mini-puffs, each representing a time interval of $\Delta t = (b-a)/n$, where a and b are start and stop times for the entire process, and n is the number of individual mini-puffs. The addition process at the observation point as n becomes large is really akin then to an integration process.

TABLE 7-B
Key contributors to one-second puff sequence results

M	4446	4448	4450	4452	4454	4457	4459	4461	4463
sigx	8.84	8.84	8.84	8.84	8.84	8.84	8.84	8.84	8.84
sigy	8.84	8.84	8.84	8.84	8.84	8.84	8.84	8.84	8.84
sigz	5.85	5.85	5.85	5.85	5.85	5.85	5.85	5.85	5.85
u	2.237	2.237	2.237	2.237	2.237	2.237	2.237	2.237	2.237
H	0	0	0	0	0	0	0	0	0
x	106.68	106.68	106.68	106.68	106.68	106.68	106.68	106.68	106.68
y	12.192	12.192	12.192	12.192	12.192	12.192	12.192	12.192	12.192
z	0	0	0	0	0	0	0	0	0
t	51.5	50.5	49.5	48.5	47.5	46.5	45.5	44.5	43.5
x-ut	-8.525	-6.289	-4.052	-1.815	0.422	2.660	4.897	7.134	9.371
coef	0.618	0.618	0.618	0.618	0.619	0.619	0.619	0.620	0.620
xterm	0.628	0.776	0.900	0.979	0.999	0.956	0.858	0.722	0.570
yterm	0.386	0.386	0.386	0.386	0.386	0.386	0.386	0.386	0.386
1st zterm	1	1	1	1	1	1	1	1	1
2nd zterm	1	1	1	1	1	1	1	1	1
sum zterms	2	2	2	2	2	2	2	2	2
	0.300	0.371	0.430	0.468	0.477	0.457	0.410	0.346	0.273

TOTAL CONCENTRATION = 4.73

Use this idea to develop a general formula for the concentration $C(x, y, z, T)$ resulting from a finite duration release on the time interval from a to b whose rate is described by a function $m(t)$. You may assume for simplicity that $T \geq b$. (Hint: this is not as imposing as it sounds if you keep good track of terms, just as in the numerical example above.) You may leave your answer in the form of an integral.)

This same kind of integration process was used in Chapter 6 to calculate concentrations from a distributed source of diffusing mass.

An arbitrary interval around t of length Δt accounts for a mass input of approximately $m(t)\Delta t$. Its contribution to the total concentration at a point (x, y, z) and time T is just given by the puff equation:

$$C_{\text{partial}} = \frac{m(t)\Delta t}{(2\pi)^{3/2}\sigma_x\sigma_y\sigma_z} \left[e^{-\frac{(x-u(T-t))^2}{2\sigma_x^2}} \right] \left[e^{-\frac{y^2}{2\sigma_y^2}} \right] \left[e^{-\frac{(z-H)^2}{2\sigma_z^2}} + e^{-\frac{(z+H)^2}{2\sigma_z^2}} \right]$$

where the exponent on the x term is precisely what we implicitly used in the previous exercise. The total concentration is then given by the limit of the sum of such contributions from over the entire interval $[a, b]$, which is just the integral:

$$C(x, y, z, T) = \int_a^b \frac{m(t)}{(2\pi)^{3/2}\sigma_x\sigma_y\sigma_z} \left[e^{-\frac{(x-u(T-t))^2}{2\sigma_x^2}} \right] \left[e^{-\frac{y^2}{2\sigma_y^2}} \right] \left[e^{-\frac{(z-H)^2}{2\sigma_z^2}} + e^{-\frac{(z+H)^2}{2\sigma_z^2}} \right] dt.$$

This is not as complex as it seems since many of the terms do not depend on the variable of integration t. Thus we can write:

$$C(x, y, z, T) = \frac{1}{(2\pi)^{3/2}\sigma_x\sigma_y\sigma_z} \left[e^{-\frac{y^2}{2\sigma_y^2}} \right] \left[e^{-\frac{(z-H)^2}{2\sigma_z^2}} + e^{-\frac{(z+H)^2}{2\sigma_z^2}} \right] \int_a^b m(t) \left[e^{-\frac{(x-u(T-t))^2}{2\sigma_x^2}} \right] dt.$$

For constant source functions $m(t)$, this factor also can be removed from the integral, and it is easy to see that by an obvious change of variables it can be written in terms of the classic error function, $\text{erf}(z)$. (This idea has been introduced previously and will not be pursued further here.)

17.   Use your integral representation from the previous exercise, together with a numerical integration routine in a standard math package, to solve Exercise 15.

The function we want to evaluate in this particular case is:

$$C(x,y,z,T) = \frac{1}{(2\pi)^{3/2}\sigma_x\sigma_y\sigma_z}\left[e^{-\frac{y^2}{2\sigma_y^2}}\right]\left[e^{-\frac{(z-H)^2}{2\sigma_z^2}} + e^{-\frac{(z+H)^2}{2\sigma_z^2}}\right]\int_a^b m(t)\left[e^{-\frac{(x-u(T-t))^2}{2\sigma_x^2}}\right]dt.$$

The coefficient of the integral (including the initial quotient as well as the y and z terms) has been calculated earlier and is reported in Table 7-A, although the part labeled "coef" there has M factored in (in grams), which needs to be divided out. In addition, $m(t)$ needs to be converted to grams per second by multiplying by the natural conversion factors. The result is the integration problem:

$$C(x,y,z,T) = (1.39\times 10^{-4})(.386)(2)\int_0^{1800}\left(100 + \frac{5t}{18}\right)\left(\frac{454.55}{60}\right)\left[e^{-\frac{(106.68-2.237(1800-t))^2}{2(8.84)^2}}\right]dt$$

$$= (8.1295\times 10^{-4})\int_0^{1800}\left(100 + \frac{5t}{18}\right)e^{-\frac{(2.237t-3919.92)^2}{156.29}}dt$$

which is found by a standard integration routine to have the value 4.725. This agrees very well with the result in Exercise 15 that we obtained when we finally broke the interval down into very small subintervals of one-second length each. Obviously the integration approach represents a desirable framework for incorporating such calculations into a modeling package, even more so in cases when, as described at the end of the previous solution, it can be transformed into a standard integral function like $\text{erf}(x)$ for which evaluation schemes already are available in most math packages.

18.   Solve the problem in Exercise 15 using your modeling package, and compare with the results obtained in that exercise and in Exercise 17.

ARCHIE does not treat time-varying vapor evolution rates, but we can approximate the situation by using the vapor evolution rate from the most influential part of the time interval. Since the travel time with the wind from the pool to the receptor is about 48 seconds, we will use the rate corresponding to a time of 30 minutes less 48 seconds. This rate is 586.7 lb/min, and this is the system of units ARCHIE will request. We will also need to input a release duration. Keeping in mind that the most influential releases are those quite close to the above time of 30 minutes less 48 seconds, we will assume a release of 96 seconds to simulate a symmetric interval around the point. (This is only one of several rationales, and one would need to study the detailed structure of ARCHIE to decide on the most logical. That is beyond our scope here. In any case, the ARCHIE results are not terribly sensitive to this choice.) With these values, the toxic vapor submodel of ARCHIE can be used to get a concentration profile, which, by interpolation to our receptor location, gives a maximum concentration there of 7,430 ppm, which converts to 16 g/m^3. This differs from the earlier results by about half an order of magnitude.

It would no doubt be interesting to investigate further how the models approach this problem differently, but that is beyond our scope. However, it is indeed common for these kinds of differences to occur, and often they result from different levels of conservatism associated with different modeling approaches. It is important to keep in mind that one cannot judge the value of a model without an understanding of the specific purposes for which it is going to be used.

19.   Discuss conceptually the use of the mini-puff concept for dealing with a "distributed source," meaning one whose spatial extent is such that it would be too inaccurate to model as a point source.

This is similar to the concept in Exercise 16, above, and to Exercise 18 in Section 6-4, where a diffusion problem with a spatially distributed source was modeled. If the receptor is at the space-time point (X,Y,Z,T), say, then each spatial location in the pool at each time can be thought of as the source of a "mini-puff," the sum or superposition of which will give the total concentration at the receptor. If the vapor emission rate is $m(x,y,t)$ in units of mass per unit pool area per time, then the resulting concentration at this receptor would be:

$$C(X, Y, Z, T) = \iiint_{x,y,t} \frac{m(x,y,t)}{(2\pi)^{3/2}\sigma_x\sigma_y\sigma_z} \times$$

$$\left[e^{-\frac{(X-(x-u(T-t)))^2}{2\sigma_x^2}} \right] \left[e^{-\frac{(Y-y)^2}{2\sigma_y^2}} \right] \left[e^{-\frac{(Z-H)^2}{2\sigma_z^2}} + e^{-\frac{(Z+H)^2}{2\sigma_z^2}} \right] dx\,dy\,dt.$$

Note that some terms can be taken right out of this integral; and, in addition, it is subject to simplifying transformations in the most common case of constant m and equal dispersion coefficients in the x- and y-directions.

20. Review the documentation for your modeling package and summarize concisely, in the framework described above, how it deals with clouds of heavy gases.

For toxic gas calculations, ARCHIE uses a Gaussian dispersion scheme. This is based on test calculations that showed that hazard distances are generally longer (and hence more conservative) using this approach. For flammable vapor calculations, ARCHIE determines whether a problem has a heavy gas nature. If it does, then it uses a simplified model that consists of curve-fitting the results of more complex heavy gas calculations for a range of materials and release conditions.

21. Consider the instantaneous release of 5,000 gallons of liquefied propane from a small pressurized storage tank ruptured by a runaway truck. Assuming that ignition does not take place as a result of impact, determine the flammable hazard zone (distance to 1/2 LFL). Assume that it is a clear night with a windspeed of 6 mph. To keep the focus on the heavy gas issue, assume that the entire mass flashes upon release. If your modeling package chooses a modeling approach for the vapor cloud automatically, what choice does it make? If you have the option of specifying whether to apply either heavy gas or Gaussian models, apply both and compare the results.

Using the specific gravity of propane ($\approx .55$), the source consists of 22,940 pounds, which we assume is released in a nominal one-minute period, the normal ARCHIE framework for an instantaneous release. The flammable vapor submodel within ARCHIE automatically uses a heavy gas model in this case and calculates a hazard distance of 1,278 feet to a 1/2 LFL concentration. We can simulate the corresponding calculation using a Gaussian model by using the toxic hazard submodel within ARCHIE. The 1/2 LFL concentration corresponds to 1.1% by volume, which is the same as 1.1% of the number of molecules, which is the same as 11,000 ppm. Using this last value as an artificial toxic vapor limit for propane, the model calculates a hazard distance of 5,007 feet, roughly four times the result obtained with the heavy gas model.

7.5 General Comments and Guide to Further Information

(No exercises.)